Game Theory in the Social Sciences

Individuals, firms, governments and nations behave strategically, for good and bad. Over the last few decades, game theory has been constructed and progressively refined to become the major tool used by social scientists to understand, predict and regulate strategic interaction among agents who often have conflicting interests. In the surprisingly anodyne jargon of the theory, they 'play games'. This book offers an introduction to the basic tools of game theory and an overview of a number of applications to real-world cases, covering the areas of economics, politics and international relations. Each chapter is accompanied by some suggestions for further reading.

The book begins with an outline of the history of game theory, whose early achievements were driven by the Cold War. Then, the definitions of games, strategies and solution concepts are illustrated as informally as possible, accompanying each of them with intuitive explanations. The remainder of the volume is entirely devoted to lay out examples related to economic issues (technical progress, advertising, cartels, monetary and fiscal policy) and a number of problems related to politics (voting and electoral campaigns) and international relations and international political economy (such as the Cold War, free trade versus protectionism, and openness versus security). The last two chapters deal, respectively, with the role of information and the strategic manipulation thereof, and cooperation and bargaining in games where agents may leave aside their selfish attitude and approach each other in a somewhat different mood. These two chapters also offer appropriate examples.

The author has intentionally taken an accessible approach to the subject matter. No particular knowledge of calculus is required to follow the book, and to grasp its message. Therefore, this volume would suit undergraduates and masters students in the social sciences without a background in mathematics, as well as anyone interested in understanding relevant phenomena of contemporary economics and politics.

Luca Lambertini is Professor of Economics and Vice Dean in the Faculty of Political Science at the University of Bologna, Italy.

Game Theory in the Social Sciences

A reader-friendly guide

Luca Lambertini

Routledge
Taylor & Francis Group

LONDON AND NEW YORK

First published 2011
by Routledge
2 Park Square, Milton Park, Abingdon, Oxon, OX14 4RN

Simultaneously published in the USA and Canada
by Routledge
711 Third Avenue, New York, NY 10017

Routledge is an imprint of the Taylor & Francis Group

Typeset in Times New Roman by Sunrise Setting Ltd, Devon, United Kingdom
Printed and bound in Great Britain by the MPG Books Group

British Library Cataloguing in Publication Data
A catalogue record for this book is available from the British Library

Library of Congress Cataloging in Publication Data
A catalog record for this book has been requested

ISBN13: 978-0-415-59111-9 (hbk)
ISBN13: 978-0-415-66483-7 (pbk)
ISBN13: 978-0-203-81864-0 (ebk)

In memory of Agostino, Mario, Hugo, Corto and Tarao.
They are now sailing together into infinity.
Safely though, as Tarao is certainly the best navigator.

Contents

List of figures   ix
Preface   xi

1 The origins: a bit of history   1
   1.1   Giant steps   1
   1.2   Hidden truths?   5

2 What is a game?   11
   2.1   The structure of a game   13
   2.2   A brief taxonomy of games   14
   2.3   Alternative representations   16

3 Solving a game   21
   3.1   The maximin (or minimax) equilibrium   22
   3.2   Refinements of the Nash equilibrium   26
   3.3   Warnings   36
   3.4   Risk dominance   39
   3.5   Nash equilibrium in mixed strategies   39
   3.6   Appendix: Schrödinger's paradox   43

4 Understanding economics   47
   4.1   Industrial economics   47
   4.2   Monetary and fiscal policies   63
   4.3   Natural resources and the environment   66

5 Repeated games and collusive behaviour   69
   5.1   The prisoners' dilemma revisited   69
   5.2   Time and time discounting   70
   5.3   Finite or infinite horizon?   72
   5.4   The folk theorems   73
   5.5   The chain store paradox   82

6 Understanding politics   85
   6.1   Voting paradoxes   85

6.2   A spatial model of political competition   88
6.3   The robustness of the median voter theorem   90
6.4   Electoral campaigns   93
6.5   How about being re-elected?   94

7   Wargames   97
7.1   The battle of the Bismarck Sea   97
7.2   Overlord   99
7.3   Escalation as an all-pay auction   101
7.4   Mutually assured destruction and the Cuban missile crisis   103
7.5   The Euromissiles crisis   105
7.6   Hawks, doves and Star Wars   108

8   Trade, security and hegemony   113
8.1   International cooperation and free trade   113
8.2   Guns versus butter and the trade-off between
*   openness and security   117*
8.3   The persistence of unipolarism   119
8.4   Appendix: the game between satellites   126

9   The role of information   129
9.1   Asymmetric information   130
9.2   Incomplete information   136
9.3   Forward induction   144
9.4   Appendix: Bayes' rule   147

10   Bargaining and cooperation   149
10.1   Bargaining games: the axiomatic approach   150
10.2   Cooperative games: a matter of coalitions   154
10.3   Examples   157

Notes   165
Bibliography   173
Index   185

List of figures

1.1	The pursuit-evasion game	8
2.1	Extensive form of a 2×2 game with imperfect information	17
2.2	Extensive form of a 2×2 game with perfect information for player 2	18
2.3	Extensive form of a 2×2 game with perfect information for player 1	18
3.1	Perfect information for player 2	33
3.2	Perfect information for player 1	34
3.3	Imperfect information (simultaneous play)	35
3.4	The battle of the sexes (ladies first)	35
3.5	The stag hunt game under perfect information	36
3.6	The centipede game	37
3.7	Perfect information for player 2	43
4.1	The Hotelling segment	55
4.2	The entry game with sequential moves	59
4.3	Quality and standardization	63
5.1	Profits, consumer surplus and welfare	74
5.2	The frontier of industry profits and the cone of Pareto-superior allocations	76
6.1	The Hotelling–Downs segment	89
6.2	Announcements and credibility under sequential moves	96
7.1	Schematic map of the battle of the Bismarck Sea	98
7.2	The Overlord game in extensive form with perfect information for the Germans	101
7.3	The extensive form of the Cuban missile crisis game	104
7.4	The Euromissiles game in extensive form	108
8.1	The extensive form of Matrix 8.10	127
9.1	A forward induction game	145
10.1	The feasible set	151
10.2	The Nash bargaining solution	152
10.3	Example: the linear frontier case in duopoly	153
10.4	Sharing the defence burden: the benchmark case	158
10.5	The core and Shapley value of the benchmark case	159
10.6	A triangular world	160

Preface

I remember when rock was young,
me and Suzie had so much fun.

Sir Elton John,
Crocodile rock

I've always cherished the idea of becoming a story teller. To tell the bare truth, when I was a teenager, I also dreamed of becoming a fighter pilot or a starship pilot – legitimate and possibly quite common aspirations for a youngster. It goes without saying that what I ended up doing for a living is neither the first nor the second. Having grown much older, and having become a mathematical economist in the meantime, I am about to tell the closest substitute to a story that I can possibly think of. It's a tale of many people's endeavours to uncover a structure in the seemingly inextricable net of social interactions, and to give this structure a sense. It is indeed an amazing story, and I hope you will enjoy reading it. As to holding in my hands the throttle and stick of a jet aircraft or starship, well, I will have to wait for a second chance in another life.

Game theory is a branch of mathematics. As such, it is just an instrument, although, unlike much of the rest of mathematics, it is a very *specific* one. It has been constructed to describe and understand social situations. Hence, this should be a topic of interest to the public at large, as it tackles issues that most people, to some extent, are familiar with, although they may have never thought about them as the subject of a formal investigation.

The task of choosing one's own best behaviour and simultaneously figuring out the behaviour of others is something we are well acquainted with in everyday life. This holds true not only for households, which have to decide whether and how much to invest in the education of their children, but also for firms, which have to figure out whether to carry out research and development (R&D) projects for new products or technologies, and similarly for nations facing the perspective of engaging in an arms race. The labels and objects may change, but these examples indeed share the same underlying structure, whose essential features are summarized in a single question that this household, firm, or nation has to answer: 'What if I don't do it, while others do?' The answer shapes the resulting strategy that the agent is going to adopt. Throughout the last quarter of a century, game theory has become the major toolkit for our understanding of strategic interplay among individuals, firms, political parties and nations. Indeed, the issue that triggered its initial development was how to fight and ultimately win the Cold War without getting to a real showdown, not how to understand firms' behaviour. For this very reason, most of the theory's early stages and achievements remained under cover for a long time, as a sort of *X-Files*. This volume offers a concise historical reconstruction of the development of the theory, followed by a friendly introduction to its instruments, and applications to the broad range of the social sciences, using as little calculus as possible, so as to enable you, the reader, to intuitively grasp the essence of economic and political issues.

Aims and style

> If you can't explain it simply,
> you don't understand it well enough.
>
> Albert Einstein

Being no more than an introduction, the book can be used as a textbook for undergraduates in any BA course in social studies (in particular, politics and economics). However, my main objective was to write a book where mathematics would not scare off readers potentially interested in grasping the flavour of a branch of applied mathematics that has become surprisingly popular over the last decade or so. Accordingly, I have tried to get around the usual obstacle of calculus, with very very few exceptions, which can be skipped without hindering the comprehension of the subjects being treated (as we usually say, 'a figure is worth 1000 words, while any equation cuts sales by 50 per cent').

Moreover, although all chapters (in particular Chapters 5–10) contain several examples and applications, they are far from being exhaustive surveys of each of the topics they deal with. For those interested in going deeper in any direction, at the end of each chapter I have included a paragraph containing some further reading. These appear in the bibliography at the end of the book, together with those references that are explicitly mentioned in the text chapters. Additionally, some of the suggested reading is not specifically connected with the technical issues treated in a chapter; rather, it has to do with its general mood and its links with other areas and disciplines, loosely related with the tale emerging from the chapter itself.

I have decided to avoid an impersonal style of writing throughout the book. I will use 'I', 'you' and 'we', in different passages, depending on what exactly each passage is about. This choice is motivated by my intention to talk to you and involve you in the reconstruction, interpretation and understanding of the facts – belonging either to the discipline of game theory or to the real world – as much as possible, although obviously I won't get any feedback while you read through. But I hope this style will help you to feel more at home.

Structure of the book

> Do not worry about your difficulties in mathematics.
> I can assure you mine are still greater.
>
> Albert Einstein

The first chapter outlines the origins and evolution of game theory, giving also some anecdotal evidence about the lives of some of the most relevant characters (Zermelo, von Neumann and Nash, *inter alia*) and the initial impulse to the development of the theory at the outset of the Cold War at the RAND Corporation.

The second chapter offers a layout of the basics of game theory: definitions, taxonomies of games, different representations (strategic form versus extensive form), without using calculus at all, just figures and graphs accompanying intuitive verbal arguments. It will also spell out an intuitive explanation of what is meant by 'rationality' and 'maximization', using some stylized examples as a warning against the indiscriminate use of these concepts.

In Chapter 3, you will find a survey of the main equilibrium concepts for non-cooperative games: the Nash equilibrium, dominant strategies, the focal point and subgame perfection. This material, I stress once again, is treated as informally as possible. Famous examples (the prisoners' dilemma, the game of sexes, coordination games) are introduced here for illustrative purposes as well as for future reference.

Having outlined the historical background and the reference framework of the discipline, the rest of the book is mainly concerned with the description and comprehension of real-world phenomena.

Chapter 4 deals with games in micro/macroeconomics, market games, public and environmental economics: price competition with homogeneous or differentiated goods, product and process innovation, advertising, taxation, pollution, entry and entry barriers, etc. These examples, wherever appropriate, are also discussed in terms of policy implications.

Chapter 5 contains a simple introduction to repeated games, using the so-called *folk theorem* in order to illustrate the arising of implicit cooperation. Repeated games pop up again in later chapters to investigate specific issues, such as the choice between protectionism and free trade in Chapter 8.

Chapter 6 is for politics. This is a classical field that is suited to applications of game theory. The paradox of voting is illustrated, with reference to a famous real-world case involving Jimmy Carter, Ronald Reagan and Gerald Ford in the 1970s, and games where parties strategically choose their respective platforms to win elections. Additionally, the prisoners' dilemma generated by investing in electoral campaigns is investigated, being formally equivalent to a game of advertising appearing in Chapter 4.

Then, Chapter 7 recounts the early views on game theory as an instrument to learn how to win conflicts, from early examples related to World War II to Cold War episodes (such as the Cuban missile crisis and the Euromissiles one).

The material included in Chapter 8 is strongly related to what is contained in the previous chapter, but deserves a space of its own as it covers the field of international political economy. Here I dwell upon issues like the choice between autarky (or protectionism) and trade (illustrating the different views that characterize the new trade theory and political science in this respect), the trade-off between openness and security, as well as hegemony and unipolarism.

In Chapter 9, I discuss the relevance of information, to see what happens if the latter is either asymmetric or incomplete, i.e., if some players are less informed than others in the same game, about some relevant piece of the game itself. In doing so, some of the games/stories appearing in previous chapters are revisited, along with famous games that are milestones in the development of game theory, such as Akerlof's famous 'market for lemons', to mention but one.

The last chapter is somewhat (although not completely) self-contained, as it illustrates the essential features of bargaining and cooperative games and the related solution concepts (basically, the Nash bargaining solution) in order to examine some real-world stories like the problem of mutual defence, international negotiations and environmental issues (the exploitation of natural resources, pollution and global warming) as cooperative set-ups.

Acknowledgements

> So there ain't nothing more to write about,
> and I am wrotten glad of it, because
> if I'd'a' knowed what a trouble it was
> to make a book, I wouldn't 'a tackle it,
> and ain't a-going to no more.
>
> Mark Twain,
> *The adventures of Huckleberry Finn*

The author's efforts are, *per se*, necessary but by no means sufficient to deliver a book. The contributions of many people, friends, colleagues and students who have read, discussed or experimented with some or all of the contents of this volume have helped me a great deal to get this material into its final shape over the last few years. My warmest thanks (but obviously not the responsibility, which remains with me only) go to Emanuele Bacchiega, Tiziano Bonazzi, Marco Cesa, Roberto Cellini, Luca Colombo, Davide Dragone, Giampiero Giacomello, Erik Jones, Piero Ignazi, Paola Labrecciosa, George Leitmann, Luigi Luini, Andrea Mantovani, Massimo Marinacci, Enrico Minelli, Raimondello Orsini, Arsen Palestini, Angelo Panebianco, Michael Plummer, Gino Segrè, students at the Paul H. Nitze School of Advanced International Studies, Johns Hopkins University, Bologna Center, and those at the Faculty of Economics of Bologna and the Faculties of Social Studies of Bologna and Forlì. And, of course, I am also grateful to the Routledge/Taylor & Francis team, Simon Holt, Rob Langham and Tom Sutton, as well as four anonymous reviewers who examined the project in its early stage.

   Last, but not least, I thank Monica, not only because she pushed me to finalize a work that had remained stored in the back of my mind for quite a while but, above all, because the best strategy you can figure out in a lifetime is to find someone you really love and then realize that this someone, strange as it may seem, loves you back. Eventually, that is what it all boils down to.

Luca Lambertini
University of Bologna, Italy

November 2010

1 The origins: a bit of history

If you really want to hear about it,
the first thing you'll probably want
to know is where I was born and
what my lousy childhood was like

J. D. Salinger,
The catcher in the rye

The *incipit* of the *Catcher* perfectly fits the essence of this chapter. Very much like Holden Caulfield – but clearly not in the form of a first-person narrative – I am setting out to tell you a story unveiling the creation and development of what is now a well-established sub-discipline of applied mathematics, and the backbone of modern social sciences.

Nowadays, game theory provides the essential analytical instruments that virtually any-one professionally carrying out theoretical research in the social sciences has to master. Furthermore, predictions and normative indications produced by theoretical models based upon game theory represent the basis for policy analysis and empirical research. This is true in particular for economics, but largely analogous considerations apply for politics and the social sciences as a whole. That is to say, the current state of our profession as social scientists is dramatically different from what it was no more than a quarter of a century ago. The work of those who have contributed to build up game theory has changed the way we look at social interactions of any type, once and for all. And yet, if one interviews a sample of students drawn from schools of economics, politics and sociology – and it doesn't really matter whether at the undergraduate or graduate level – it clearly emerges that they have a very vague idea – if any at all – of the intellectual process that has generated those instruments they are being taught to use.

In order to allow you to come to grips with the essential history of game theory, I am setting out to accomplish a twofold task. One side of it consists in outlining the fundamental steps of the theory itself, without delving too much into the technicalities involved. In summary, this requires a short reconstruction of the development of equilibrium concepts that will be properly reviewed and then applied several times in the remainder of the book. The other side looks, at least to me, somewhat more novel and intriguing, as it requires a reconstruction of a number of relevant facts that affected the circulation of ideas and the deployment of the theory, as well as its utilization in the social sciences, for decades.

1.1 Giant steps

First of all, why *game theory*? A more appropriate label would certainly be the *theory of strategic interaction*, as this would transmit a correct idea of what it is. The reason for the choice of a label that immediately came into use can be traced back to the interest of our fore-runners in parlour games such as chess, backgammon and cards.[1] Indeed, what is commonly recognized as the first formal result in game theory is

- Zermelo's theorem (Zermelo, 1913), which refers to the outcome of a chess game. A widely accepted view is that the theorem states that either the 'white' player or the 'black' player wins, or they draw. To be more precise, the theorem does not identify the correct strategies for players, and does not predict the outcome of a chess game (see Aumann, 1987, p. 461).[2]

Not so exciting, you say? Well, I cannot raise any particular objection, except to invite you to consider that Zermelo's theorem relies on the backward induction principle, which will pop up again several times throughout this book and plays a fundamental role in game theory. Roughly speaking, the idea is that, in a chess game with alternate moves, players must be able to correctly recall the exact sequence of moves up to any point, and use this information rationally to solve the game itself. Aside from this, the object of Zermelo's attention is indeed nothing more than a parlour game. The second fundamental step of the theory is, however, far more relevant in itself as well as in terms of its potential applications and extensions. This is

- John von Neumann's minimax theorem (von Neumann, 1928), which is the first solution concept for non-cooperative games, although applicable only to constant-sum (or zero-sum) games, i.e., those where statements such as 'I win, you lose' hold systematically. Indeed, as the original paper was published in German (like Zermelo's), the relevance of the minimax equilibrium concept became apparent only after the publication of the volume *Theory of Games and Economic Behavior* (von Neumann and Morgenstern, 1944). This publication considerably enlarged the interest of the profession in game theory, in addition to building up the foundations of modern microeconomics by formalizing the expected-utility theory dating back to Ramsey (1931).

What occurred between 1928 and 1944 deserves closer inspection. You may be familiar with the main character played by Peter Sellers in Stanley Kubrick's famous movie *Dr Strangelove or: How I Learned to Stop Worrying and Love the Bomb*. Dr Strangelove is a scientist constrained in a wheelchair and characterized by a strong German accent, as well as a definite inclination towards carrying out a nuclear first strike upon the USSR. This figure, they say, was inspired by some of John von Neumann's traits, mixed up with some others belonging to Edward Teller. The two men actually had a few important things in common, in particular the fact of both being (i) Hungarian-born scientists, (ii) involved in the US programme for the development of nuclear weapons that started during World War II with the Manhattan Project, and (iii) definitely anti-communist. They were both Jewish, which suffices to reveal how misplaced is another feature of Dr Strangelove, that is, his right hand jumping up repeatedly and uncontrollably in the Nazi salute. This, in the movie, qualifies our doctor as a Nazi scientist arguably brought to the USA by OSS (that later became the Central Intelligence Agency) and most likely involved in the armaments race engaged with the Soviet Union. Well, this character cannot really fit the portrait of John von Neumann, except for the fact that the latter was constrained for years in a wheelchair by the cancer that eventually killed him.[3] He was born Janos Neumann in 1903 in the Hapsburg Empire. His father, a banker in Vienna, was made a Knight of the Empire by Emperor Franz Josef in

1913, whereby the noble prefix. Already a gigantic figure for his extraordinary achievements in pure and applied mathematics and in physics,[4] he was convinced by Oskar Morgenstern, an Austrian economist, to leave Europe and reach Princeton in 1930, joining the Institute of Advanced Studies that had been created to host a large number of illustrious scientists who were abandoning the old continent in those years (among them, Albert Einstein, Enrico Fermi and Kurt Gödel, to mention only a few). As an economist, Morgenstern had intuitively understood the potential associated with the newborn game theory and strived to convince a comparatively reluctant von Neumann to join their resources into a venture that ultimately yielded *Theory of Games and Economic Behavior* towards the end of the war.[5] However, after the publication of the book, the attention of von Neumann was driven away from developing game theory any further. In this respect, he confined himself to a series of regular seminars at the Institute, being more than busy on several other fronts unrelated to the social sciences.

His seminars, and the lack of a solution concept for variable-sum games, attracted John Nash to Princeton. In 1949, Nash, a young graduate from Carnegie Mellon, started working on his doctoral dissertation under the supervision of Albert Tucker. Unlike what is represented in the movie *A Beautiful Mind*,[6] Princeton was *the* place for game theory at the time, with many smart young mathematicians of the same calibre as John Nash putting their best efforts into the development of the theory.

One beautiful day, Nash stepped into the office of von Neumann with a piece of paper containing what we know now as

- the Nash equilibrium (Nash, 1950a, 1951). This solution concept can be used to solve any zero- or non-zero-sum non-cooperative game and is *de facto* a generalization of von Neumann's minimax equilibrium.

Von Neumann reacted without much enthusiasm, saying that Nash's theorem was nothing more than a straightforward application of the fixed-point theorem (Brouwer, 1910; Kakutani, 1941). Notwithstanding that, being subsequently encouraged by Tucker and others, Nash got his result quickly published.[7]

Nash decided to go straight to von Neumann's room because Tucker was not in Princeton, but somewhere else. Where? At RAND Corporation, located in Santa Monica, California. This reference to RAND is not purely anecdotal, since this institution has played a crucial role in our story. I will come back to this point below.

Over the last decades, a lively debate has taken place about the existence and identity of forerunners of the Nash equilibrium concept. Perhaps I should immediately spell out my personal opinion, as I believe this issue is not soundly posed. Leonard (1994), with the aim of showing that Nash cannot be nested into a parental line that goes ultimately back to Cournot (1838), argues against Aumann's position (Aumann, 1985), according to which the Nash equilibrium concept had always existed in the economic literature, and all that Nash has done is nothing but shaping it into its final and most general formulation. As I see it, Aumann never literally meant to single out Cournot as a forerunner of Nash. Rather, Aumann may have taken the Platonic view (which, by the way, is common to many mathematicians), maintaining that mathematical concepts have a life of their own and from time to time some researchers attain them in a more or less accurate and general form. Nash would be a representative example of the first type, while Cournot, Edgeworth and others are examples of the second type.

Be that as it may, over the years it rapidly became clear to the profession that a game could produce a plethora of Nash equilibria. This gave rise to an intensive research effort aimed at defining the so-called refinements of the Nash equilibrium concept, such as trembling hand perfection and iterated dominance. To the aim of the present chapter, it will suffice to draw the reader's attention to one such refinement:

- the subgame perfect equilibrium by backward induction (Selten, 1965, 1975). The concept of subgame perfection (or perfectness) allows one to eliminate Nash equilibria based upon non-credible strategies in non-cooperative multistage games characterized by perfect information, and it is probably the most widely used equilibrium concept in the current game-theoretic literature in the social sciences. The subgame perfect solution by backward induction is closely associated with the extensive form representation of games also known as the *Kuhn tree* (Kuhn, 1953). Hence, one could succinctly say that

$$\text{Zermelo} + \text{Nash} + \text{Khun} = \text{Selten}$$

All of the aforementioned solution concepts apply to non-cooperative (possibly multistage) games characterized by complete information (no matter whether it is perfect or not). The next step deals instead with the issue of solving non-cooperative games under incomplete information, performed via

- the Bayes–Nash perfect equilibrium (Harsanyi, 1967/68). This solution concept applies to games of incomplete information, showing that they can be treated as games of complete but imperfect information.

A further development, which mainly relies upon the concept of subgame perfection by backward induction, concerns the analysis of repeated games (or supergames), where the constituent game is a prisoners' dilemma affected by a free-riding incentive that, in the one-shot case, induces the emergence of a Pareto-inefficient Nash equilibrium in (at least weakly) dominant strategies. The main result of this particular stream of research is known as

- the *folk theorem*, a label accounting for the fact that it is unclear who coined it first – although Robert Aumann is an obvious candidate. The earliest formulation of it made use of the infinite reversion to minimax strategies as a deterrence against deviations from the collusive path. The first folk theorem in written form is Friedman's (1971), using the Nash reversion, with many follow-ups refining the theorem itself (see Axelrod, 1981; Abreu, 1986; Fudenberg and Maskin, 1986, *inter alia*) through the introduction of (optimal) punishments that are more severe than the Nash behaviour.

So much about non-cooperative games. What if, instead, players can rely on binding – or enforceable – agreements? What if they realize that they can improve upon the non-cooperative outcome through successful negotiations based upon credible and commonly accepted instruments? Parallel to (and intertwined with) the development of non-cooperative game theory, the areas of cooperative and bargaining games were also building up a framework of their own. A remarkable aspect of this side of game theory is that John Nash, via the so-called *Nash programme*, introduced the idea that cooperative games could be treated as a special case of non-cooperative ones (Nash, 1951). Nash's hint ultimately flourished into the fundamental contributions of Harsanyi and Selten (1972) and Rubinstein (1982), and also led the profession to consider the possibility of generating quasi-cooperative outcomes through the implicitly collusive behaviour based upon the repetition of a given constituent game, which takes us to the aforementioned folk theorem.

The first solution concept for cooperative games is the *core*, whose earliest (although perhaps somewhat implicit) use can be found in Edgeworth (1881), while its formalization has been performed by Luce and Raiffa (1957) and Gillies (1959). A refined version, already present in the literature before the latter works, is the concept of *stable set* by von Neumann and Morgenstern (1944). The cornerstone of cooperative game theory – if one should name a single instrument – is most probably the *Shapley value* (Shapley, 1953), while the backbone of bargaining games – where agents are supposed to negotiate gains against what they can non-cooperatively attain under disagreement – is provided by the *Nash bargaining solution* (Nash, 1950b) and the *Kalai–Smorodinsky solution* (Kalai and Smorodinsky, 1975).

This brief and intentionally incomplete overview summarizes the official history of the development of game theory in terms of its toolkit.[8] What can we say about the unraveling of applications? Was there a Trojan horse that facilitated the adoption of game theory as a dominant and widespread method of analysis in the social sciences?

1.2   Hidden truths?

> I don't think the intelligence reports are all that hot.
> Some days I get more out of the New York Times.
>
> John F. Kennedy

The commonly accepted view holds that game theory was first adopted by industrial economics in the second half of the 1970s, to become later the standard method in virtually any areas of economics. This interpretation seems to be confirmed by the casual observation of the increasing portion of game-theoretic research appearing in all economic journals during the last forty years or so. According to the same view, game theory entered the neighbouring social sciences (in particular, political science) only after having acquired a rock-solid position within economics. Hence, a widespread opinion among economists is that we may claim some sort of priority or merit for the adoption of game theory and the resulting conceptual 'revolution'.

The essential question is: Why did it take us so long to start using game theory extensively after the publication of Nash's main works in the early 1950s? After all, a quarter of a century is a long time. Relatedly, why did game theory start to percolate into economics via industrial economics, instead of some other sub-discipline?

The acquired wisdom, in this respect, is that in those years the best minds in mathematical economics had turned their attention to the construction of something economics had been wanting for a long time, namely, the theory of general equilibrium *à la* Arrow and Debreu (1954).[9] This is one of the greatest achievements, perhaps *the* greatest one, in the entire field of microeconomics. Certainly, it is the most elegant model that we have ever built up. Yet, it has very little, if any, practical purpose, as it describes a 'Paradise Lost' where all markets are perfectly competitive and prices are equal to marginal costs everywhere. It is, of course, a benchmark against which we may comparatively assess the inefficiency of real-world markets and use this information to design regulatory measures, though it remains hopelessly far away from reality. But still, the idea that the efforts to build general equilibrium theory absorbed resources that otherwise could have been used to accelerate the adoption of game theory should be considered with caution, since game theory itself rapidly found fertile applications in the field of general equilibrium theory (see Shubik, 1959; Aumann, 1964; Gabszewicz and Vial, 1972).

An alternative and complementary explanation can be found in Martin (1993, 2002, 2004), which interprets the entry of game theory into industrial economics as a reaction of part of this discipline to the Chicago School's view that dominated the 1950s and 1960s, according to which 'two is enough for competition'. This sounds pretty much like 'it takes two to tango', but the first impression may not be correct. Let me clarify this point. Consider a market with a very small number of firms, say, just two. If (and this is a crucial 'if') they supply similar goods and are extremely aggressive, the resulting degree of market competition will drive prices down, very close to marginal cost. As we shall see in Chapter 4, where I will come back extensively to this problem, this situation can be very productively analysed as a non-cooperative game. Here, I will confine myself to saying that, in the 1950s, this informal argument was used to convey the message that intensive regulation was not an issue and the State should have kept itself off the market, as the latter possessed a workable self-regulating device. Refusing this position, other industrial economists started using game theory to produce robust arguments against the Chicago view – one intuitive example being the harmful cartel behaviour that the same two firms could decide to turn to as soon as they realize that price wars ultimately destroy the profits of both. These facts triggered a process that is still ongoing and ended up involving also politics, sociology, psychology, etc.

Well, judging from the information that surfaces from the mainstream economic literature and a casual observation of what happened in the other social sciences over the past 30 years, this may look like a plausible reconstruction, but what I am about to tell you will uncover a different plot that has largely (and long) remained behind the curtains.

As mentioned above, Tucker was not in Princeton when the young John Nash came up with a new notion of equilibrium for non-cooperative games. Nash's supervisor was at RAND Corporation, a think-tank created after the end of World War II with conspicuous resources previously allocated to finance the continuation of the conflict with Japan. The idea of investing them to create an instrument to fight the new Cold War has to be traced back to the US Air Force (see Nasar, 1998). The source of the name given to this institution is interesting in itself, as it is indicative of the general mood that dominated those days. Since the research activities carried out at RAND were to be kept under cover, the name should not transmit any compromising information. The choice of the name RAND is connected with the fact that its original location was the R&D plant of Douglas Aircraft at Santa Monica Airport, where the initial Project RAND was set up in December 1945 as a special contract to Douglas Aircraft. In 1947, RAND moved from the Douglas plant at Santa Monica Airport to offices in downtown Santa Monica. In May 1948, RAND was transformed into a non-profit corporation under the laws of California. As has very often been the case, the US Navy had to have an analogous toy of its own. To satisfy this need, the Office of Naval Research (ONR, Monterey, California) was founded in 1946 as an independent (non-profit) research centre, also connected with the analysis of the Cold War.[10]

The two institutions rapidly attracted many among the best researchers from all fields (mathematics, statistics, physics, engineering and the social sciences), who were hosted there to stimulate them to find a way to win the Cold War, or, had it become hot, a conventional and nuclear confrontation with the USSR. A succinct and sharp account of how those working at RAND and ONR were perceiving their duty and the rationale sustaining it clearly emerges from the words of Isaacs (1965, p. v):

> Then under the auspices of the US Air Force, RAND was concerned largely with military problems and, to us there, this syllogism seemed incontrovertible:
>
> 1　Game theory is the analysis of conflict.
> 2　Conflict analysis is the means of warfare planning.
> 3　Therefore game theory is the means of warfare planning.

On the one hand, the stimulus administered by the military provided that extraordinary sample of top-notch researchers at RAND and ONR with the appropriate incentives to put their best efforts into this venture, and this indeed yielded astonishing results. On the other hand, however, obvious secrecy requirements prevented many results from circulating freely in the international scientific community for decades, and also implied funny consequences on the perception of what exactly was the real nature of most of the (comparatively few) results that did reach the outer world. In this regard, two games can be taken as illustrative examples.

The first game is the prisoners' dilemma, which is very well known and familiar to scholars and students across the board of social sciences and also to a wider public. This game is a prototypical strip-down model commonly used to introduce undergraduates to the idea that Nash equilibria are very often inefficient outcomes. The structure of the prisoners' dilemma was conceived and carried out experimentally by Merrill Flood and Melvin Dresher in January 1950 at RAND Corporation (Flood, 1952, 1958). According to the official records of the time, the game became famous under that name when, shortly after, Albert Tucker was invited to deliver a lecture on the nature and scope of game theory to a non-technical audience at a meeting in Stanford (see Nasar, 1998 and Mirowski, 2002). Of course, it was not a matter of understanding whether either prisoner (or both) might be driven to confess their crimes to a district attorney being worried about the lack of hard evidence. Rather, the real objective of the game was the analysis of escalation in arms races, but nobody was allowed to go public and tell the bare truth about it.

An even more intriguing case is that of a game that most people may not have ever heard of. It's traditionally known as the *pursuit-evasion game*, and its relevance goes well beyond the fact that it shares with the prisoners' dilemma a similar history and fate. Just like the prisoners' dilemma, this game also has a label that has been designed to hide its real nature. To this regard, let me tell you quickly what my personal experience with it has been. When I was a young PhD student in the early 1990s, I came across the theory of dynamic games. These are games where the role of time and time discounting is explicitly accounted for. Typical applications include the analysis of firms' investments in R&D and productive capacity accumulation, the exploitation of exhaustible natural resources, the environmental effects of productive activities, and so on. Being attracted by what later would become my main field of research, I started reading avidly everything about dynamic games, and very quickly realized that a *leitmotiv* of that literature was a strange problem called the pursuit-evasion game, with entire chapters of dominant textbooks being devoted to its illustration.[11]

This game is a dynamic set-up illustrating the strategic interaction taking place between a thief (τ) and a policeman (p) who is trying to catch him. In its simplest formulation, the game unravels over a finite time horizon ending, say, at some instant T (known *a priori* to both players). The outcome is as follows. If the policeman does not reach the thief before

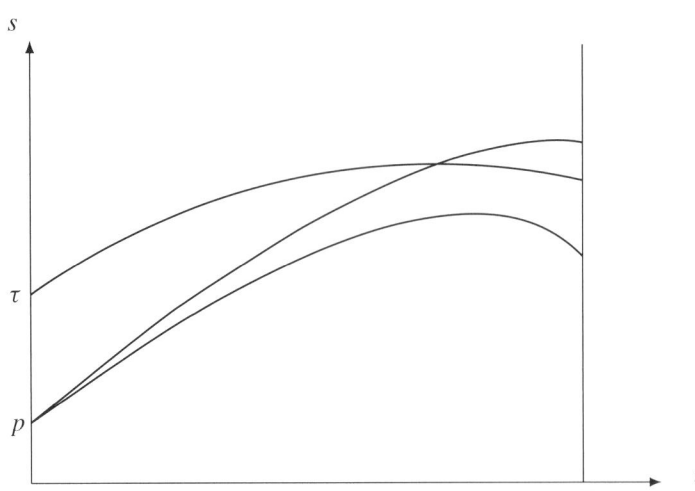

Figure 1.1 The pursuit-evasion game

the terminal date, that is, at any instant $t < T$, then the thief escapes and ultimately wins. In the opposite case, the policeman wins. The two alternative situations are illustrated in Figure 1.1, where time $t \in [0, T]$ is measured along the horizontal axis while a simplified one-dimensional space is measured on the vertical axis, and the trajectory followed by the policeman may or may not intersect that of the thief. The early sources[12] treating the pursuit-evasion game never revealed explicitly what the story was really about. As I finally came to understand later on, the subject matter of the game is the design of optimal strategies in a game where the thief is indeed a hostile intruder (e.g., an enemy aircraft or missile), possibly armed with nuclear weapons, while the policeman is the air defence of a country being attacked. Given that the intruder will reach and destroy its objective (say, the policeman's capital city) at time T, the air defence must necessarily intercept it strictly before doomsday, at any $t < T$.

Appearing under the shorter label of 'pursuit game' according to the records (cf. Breitner, 2002, p. 116), this story appeared for the first time in the RAND discussion paper P-257 written by Isaacs (1951). The pursuit-evasion game is a *differential game*. In summary, differential game theory can be viewed as a spin-off of optimal control theory, describing the dynamic behaviour of agents (or objects) in continuous-time models. As such, it is more familiar to mathematicians and control engineers than to social scientists, although its relevance for the social sciences should be self-evident. Its development rapidly became a primary activity at RAND and ONR in order to produce sophisticated instruments for the analysis of conflict situations. The same was happening also on the other side of the Iron Curtain, in the Urals, where the research in this field was driven by mathematicians like Pontryagin, Boltyanskii, Gamkrelidze, Mishchenko and many others. Secrecy, and the tough competition characterizing scientific progress during the Cold War, led to parallel work and analogous discoveries on both sides, as became clear when, with great delay, their respective achievements were officially published (Isaacs, 1973 and Pontryagin, 1966).[13] These, although altogether unrelated to 'little green men' or 'bug-eyed monsters' from outer space, might well be called the *X-Files*.

Somewhat paradoxically, the instruments of dynamic analysis have been (and largely continue to be) more familiar to macroeconomists than microeconomists or all those using the game-theoretic tools. This is confirmed by the trivial observation that the dominant models in economic growth theory (Ramsey, 1928; Solow, 1956; Swan, 1956) are much older than any publicly accessible illustrations or applications of differential game theory.

Overall, my view is that the above considerations suffice to prove two reciprocally connected facts. The first is that a generic intention of investigating economic and social issues in a broad sense is not the best candidate as the driving force behind the construction of game theory. Rather, this stream of research was motivated by the need to deal with problems specifically related to the Cold War and international relations. While it is certainly true to a certain extent that some of the researchers involved in the venture had in mind social as well as economic problems (and John Nash is certainly an example of this kind), one cannot underestimate the strength of the impulse exerted by military applications on the development of both static and differential game theory. The second fact is that the acquired interpretation shared by the vast majority of those who deal in various degrees with game theory and its applications is, in a sense, completely mistaken. That is, when reading about game theory at any level, it is very often the case that one is left totally unaware of the true genesis of the discipline, and is led to believe that the first and by a large distance – thus far – also the widest field of application has been economics.

Further reading

For more on the genesis and development of game theory, see the special issue of *History of Political Economy* on game theory (Weintraub, 1992) and Leonard (1994, 1995). A thoughtful insight into the scope and aims of game theory can be found in Aumann (1985, 1999). Interesting appraisals of the achievements of game theory are offered by van Damme and Weibull (1995) and Myerson (1999). An unconventional and intriguing view of game theory (and mathematical economics in general) is that of Mirowski (2002). For a more informal but very entertaining treatment of the matter, I refer the interested reader to Poundstone (1992), Nasar (1998) and Siegfried (2006).

2 What is a game?

Life, friends, is boring. We must not say so.
After all, the sky flashes, the great sea yearns,
we ourselves flash and yearn,
and moreover my mother told me as a boy
(repeatingly) 'Ever to confess you're bored
means you have no
Inner Resources.'

John Berryman,
Life, friends, is boring. We must not say so.

A game is a mathematical instrument that serves the purpose of formalizing strategic inter-actions among agents, the latter being not necessarily single individuals – indeed, we will examine games where players are households, firms, public institutions, nations, etc. Uncountably many social, economic and political situations, more or less familiar to all of us, are characterized by the interplay among actors with either converging or (possibly more often) conflicting interests. This prevents them from attaining what a neutral external observer would consider as their common interest.

This chapter offers a glance at the basic elements of game theory, in order to define what is the object – a 'game' – of our interest here, and what exactly one must make sure to know in order to properly construct any given game. As you can imagine, this may quickly become a distasteful overload of obscure mathematical formulas and incomprehensible def-initions, with predictably negative effects on your willingness to read on. I will do my best to avoid this unpleasant practice and its disgraceful consequence. Two examples, which I preliminarily illustrate without going into any technical details – whose exact nature will be investigated in the remainder of the chapter – precisely because I want to focus on purely intuitive aspects, will help to clarify the basics of the subject matter.

The first example is familiar to most students around the world.[1] Let's say that two Italian undergraduates, called 1 and 2, are enjoying a term at a British university thanks to some Erasmus exchange agreement. They share a little flat with a nice living room whose bow window yields a lovely view over the meadows surrounding the university campus. However, the landlord has not equipped the flat with a TV set, and our two youngsters are discussing the possibility of buying one from a local dealer. They would like a TV set priced at £200, and each of the young men evaluates the perspective of watching television comfortably at home as k. Now suppose that k is higher than £100 – that is, higher than 50 per cent of the price – and each student's available budget is £200. Under these conditions, each of them could individually decide to buy the TV – however, he could not prevent the other

Matrix 2.1 Buying the TV is a challenging venture . . .

		Student 2	
		g	*ng*
Student 1	*g*	$k + 100; k + 100$	$k; k + 200$
	ng	$k + 200; k$	200; 200

from watching it once it is in the living room; and collective fruition has no reflections at all on the individual fruition by either one.[2] Additionally, it would be wise and convenient to share the cost evenly, thereby getting the TV and saving £100 per head. The two students are following courses in two different faculties, say, engineering and medicine. The students order the TV set on the phone, and the night before the day they are supposed to go and buy it, they establish that they will meet inside the shop in the late afternoon, on the way home from their respective faculties. Now, if both of them do show up at the shop, each of them will pay £100, while if only one goes to the shop while the other doesn't, the former will have to pay the full price. It's easy to verify that if $100 \leq k < 200$ neither of them will show up, and no TV set will ever show up in their living room. The reason for this mishap can be outlined by looking at Matrix 2.1, where *g* and *ng* stand for *to go* and *not to go* to the shop, respectively.

Since $k + 200 > k + 100$, if student 2 goes to the shop, the best thing for student 1 to do is to go home and wait for 2 to come back with the TV. Otherwise, if 2 goes home, the best choice that 1 can take depends on how much he likes watching the TV, against the alternative of saving £200. Indeed, if $200 \leq k < 300$, 1 will opt to save £200. By symmetry, student 2 will get to the same conclusion. Accordingly, if $100 \leq k < 200$, the best strategy is to step back and wait for the other to come back home with the TV. As a consequence of this reasoning, neither one nor the other will go to the shop, and they will sit comfortably in the living room, but without TV.

The second example deals with the sketch of an arms race, which was a striking feature of international relations between the USA and the USSR during the Cold War. In this game, the players are two countries that have to decide whether or not to build up nuclear weapons. In principle, both would agree that a world free of nuclear weapons is much better than one where nuclear weapons appear by the thousands here and there like as many mushrooms. Yet, as for Hamlet, the question haunting their dreams symmetrically is: If my adversary is building up its nuclear capability, why shouldn't I do the same? Being peaceful – or, at least, non-aggressive – may not be so smart a strategy in a world where hegemony and bargaining power are the direct product of the number of ballistic missiles one can deploy at any given instant, right? This problem is portrayed in Matrix 2.2, where strategies *m* and *nm* stand for *missiles* and *no missiles*, respectively.

The numbers appearing in the matrix measure the values attached by the two countries to each of the four possible situations. Of course, I have chosen these numerical values arbitrarily, but not so arbitrarily as it may seem at first sight. They have been picked in such

Matrix 2.2 . . . but an arms race is even worse

		Country 2	
		m	*nm*
Country 1	*m*	10; 10	200; 0
	nm	0; 200	100; 100

a way to reflect more or less correctly the attitude of the nuclear powers during the period stretching from the end of World War II to the late 1980s. So, the message conveyed by the above matrix is that everybody shares the view that a world without missiles is definitely preferable to any other alternative, and yet having no missiles is somewhat embarrassing if the counterpart has built an arsenal for itself. Since $10 > 0$ and $200 > 100$, both countries will strive to build up their own ballistic missiles, as this emerges as the most convenient strategy irrespective of the rival's behaviour.

These two simple stories may seem quite unrelated to each other, in terms of their respective subjects. However, they share the same underlying nature, and produce analogous outcomes. Namely, in both games, the players are fully aware that there is an obviously optimal outcome (splitting evenly the cost of a TV set, or a pleasant world without nuclear weapons). However, because of strategic interplay (and the 'logic' driving it), the outcome we may expect to observe (and, in the second case, we did observe) is far worse. In both cases, I have reconstructed the likely unravelling of events under the assumption (implicit, thus far) that agents adopt quite selfish attitudes. Of course, one could consider the alternative perspective in which agents are altruistic by nature, and therefore keen on cooperating towards the attainment of the common good. It goes without saying that this, desirable as it may be, is not generally the case. This distinction, based on the players' attitude, is a sort of iron curtain separating two worlds: one in which players are selfish, the other in which they are altruistic. Most of the book deals with the first (but the last chapter is entirely devoted to the second).

These preliminary observations suggest the presence of some recurrent features repeating themselves unchanged across several different situations belonging to many areas of our life. Formalizing these problems as games is a way of capturing this common architecture without being driven off course by specific (and more often than not, irrelevant) details.

The first step in this direction consists in understanding what is meant by 'game' in applied mathematics. Admittedly, as I have said at the outset, going through a list of definitions and formal concepts is a prerequisite that may deter you from going any further. If the present one were a standard textbook in game theory, I might agree with you – at least to some extent. But I'm warmly inviting you to proceed with confidence.

2.1 The structure of a game

In order to properly define any given game (and consequently understand what it is about), one has to know the following three essential pieces of information, which jointly define the so-called *structure* of the game.

- The full list of the players' identities, which, in principle, might be just a list $N = \{1, 2, 3, \ldots, n\}$.
- The full list of actions or strategies[3] S_i that every player can choose in the game:

$$S_i = \{s_{i1}, s_{i2}, s_{i3}, \ldots, s_{ik}\} \tag{2.1}$$

In this particular case, player i has k admissible strategies. If S_i and N are both finite (that is, ten or ten billion, but not infinitely many), the game at hand is said to be *finite*. The order according to which player i's strategies are listed in S_i is irrelevant. What matters is just that S_i must be *complete*. An admissible outcome of the game is defined as

$$s = (s_1, s_2, s_3, \ldots, s_N) \tag{2.2}$$

which is the list of the individual actions or strategies chosen by each player $i = 1, 2, 3, \ldots, n$ corresponding to the outcome s. In this case the order does matter, as s_1 identifies the behaviour of player 1, s_2 that of player 2, and so on.

- The payoff accruing to each player i corresponding to every possible outcome s of the game. This payoff is usually indicated as $\pi_i(s)$ (this habit has much to do with the fact that, in economics, the Greek letter π is commonly used to identify profits). For the sake of clarity, it is worth stressing that the payoffs can be represented by anything that matters to players: money, market shares, political power, hegemony, fame, rare butterflies or stamps, vintage hi-fi items, etc. In practice, the exact nature of π_i has to be defined in relation to the player's objective in any specific game.[4]

Throughout the book, I will use the term *strategy* to indicate a player's behaviour. However, as long as strategies are discrete (e.g., *left* versus *right*, *stop* versus *go*), I could as well use the term *action*. For future reference, it is also useful to define what is meant by a *mixed strategy*. Now we know that agent i may play a pure strategy s_{ij} as long as this is admissible, that is, s_{ij} must be an element of S_i. Alternatively, the same player may *randomize* over his/her set of pure strategies and generate a mixed strategy. What does all this mean? To understand it, consider the simplest case where randomization may operate, i.e., suppose player i's admissible pure strategies are $S_i = \{s_{i1}, s_{i2}\}$. He/she may attach probability $\mathfrak{p}$ to s_{i1} and its complement $1 - \mathfrak{p}$ to s_{i2}. The resulting combination of pure strategies generated by these two probabilistic weights is $\sigma_i = \mathfrak{p}s_{i1} + (1 - \mathfrak{p})s_{i2}$ and is called a mixed strategy. For instance, i may declare (for some reasons that I will not illustrate here) that he/she will set $\mathfrak{p} = 1 - \mathfrak{p} = 1/2$, or equivalently that each pure strategy will be selected with the same probability. To help intuition, one may think of mixed strategies as lotteries constructed on pure ones, with the *caveat* that, while in real-world lotteries probabilities are exogenous to each single agent, here the agent assigns probability endogenously to each pure strategy. How to work with mixed strategies (and why one should do it) is one of the arguments discussed in Chapter 3.

2.2   A brief taxonomy of games

Our next task is to classify the different types of games according to four commonly accepted typologies. To begin with, we can separate *cooperative* from *non-cooperative* games. In a cooperative game, players are supposed to pursue a common objective, summarized by a common payoff; on the contrary, a non-cooperative game is one in which each player is assumed to adopt a strictly self-interested behaviour, in open conflict with all other players involved in the same game.

The second classification separates *constant-sum* or *zero-sum games* from *variable-sum games*. In the former type, the overall sum of payoffs accruing to all players is constant irrespective of their behaviour, in such a way that, whatever a player may gain, he does so at the expenses of at least one of the others. Hence, constant-sum games fully embody the spirit of the Latin (and, later, Hobbesian) tag *homo homini lupus*. Or, equivalently, players know that they have to split a pie whose size is given and unmodifiable endogenously. For this very reason, constant-sum games are often referred to as *strictly competitive*. Instead, in a variable-sum game, the size of the pie to be split among players does depend endogenously on their behaviour. Hence, although it remains true that every player is in conflict with all others, any gain for an individual does not necessarily entail a loss for at least one other. To eliminate any residual doubts: while a game between two nuclear powers trying to split the globe into two areas of influence might be taken to be (but, beware, *not necessarily is*) a constant-sum game, competition between, say, DrinkThis and TasteThat is surely not, since each firm's profits, as well as their sum, do depend on strategies involving prices, marketing techniques, advertising campaigns, etc.

The third classification relies on the nature and amount of information available to each player when the latter has to select a strategy or action. In this respect, three different taxonomies can be adopted.

- A game is one of *perfect information* if every player, when choosing a strategy, knows the entire past history of the game up to that moment. Spelled out in these terms, the definition of perfect information seems intuitively clear, but it's indeed a bit tricky. Why? Because most of the games we will examine in the remainder of the book are in fact one-shot games, i.e., situations where strategic interaction takes place once and for all, without past or future developments, and calendar time doesn't play any role in them. Accordingly, it's not necessarily clear what is meant by 'past history'. An alternative (and perhaps more intuitive) definition is the following: information is perfect if a player, when adopting a strategy among those allowed to him in the game, knows the strategies adopted by rivals. If this is not the case for at least one player, then the game under consideration is characterized by *imperfect information* (note that one player, or several players, may move before or after the rest of the players involved in the game, while the latter move simultaneously).
- A game is one of *complete information* if each player knows its structure, that is, the list (or the identities) of all players, the admissible strategies S_i for all of them, and their respective payoffs π_i corresponding to every admissible outcome. If this is not true for at least one of them, then the game is characterized by *incomplete information*.
- Finally, a game is one of *symmetric information* if all players have exactly the same amount of information relevant to the solution of the game. Again, if this is not true, the game will be affected by *asymmetric information*.

It is worth stressing that, more often than not, asymmetry and incompleteness will indeed overlap, so that a game with incomplete information may frequently be one of asymmetric information as well. For instance, if DrinkThis knows the technology (and therefore the cost structure) of TasteThat, while the converse is not true, then the game between DrinkThis and TasteThat will surely feature asymmetric information, but this will go hand in hand with incomplete information because TasteThat will be unable to tell exactly what price strategies can be implemented by DrinkThis, as the latter's costs are unknown to TasteThat. To stick with this example dealing with the soft drinks industry, a game with purely incomplete (but symmetric) information is one where, for instance, DrinkThis and TasteThat know that there may be a shock on demand due to affect their profits, but neither firm knows whether the shock will occur, and if so, whether tomorrow, or next week, or next year. In a situation like this, both firms know exactly the same things, but they cannot foresee precisely what pattern that market demand is going to follow in the near future. Intuitively, the aforementioned typologies can be mixed up in any possible order, giving rise to games where information is imperfect but both symmetric and complete, or perfect, symmetric and incomplete, or anything else.

The last classification of games has to do with the role of time. In this respect, there exist two types of games: *one-shot games* on one side, and *repeated games* or *supergames* on the other. In the first case, the game is played one-off: players meet just once, there is neither past nor future, and therefore time plays no part in the story. In the second case, the constituent game (something like Matrices 2.1 and 2.2) repeats an arbitrary number of times, even infinitely many times. As we shall see in Chapter 5, time matters, so much so that repetition may indeed modify dramatically the outcome of a game. For the moment, I will confine my attention to one-shot games.

2.3　Alternative representations

On the basis of the notions listed in Section 2.1, we may define a generic game G in what is usually called the *normal form* $G \equiv \langle N, S_i, \pi_i(s) \rangle$. This synthetic representation of a game captures the idea that the game itself is well defined if its structure is known. However, this way of describing the essential features of games is somewhat less than intuitive, and I will use it only when strictly necessary to spell out theorems or definitions. Fortunately enough, this will not happen often in the book, as there are alternative – and more friendly – methods of describing games. In fact, there are two standard – but not equivalent – ways of visualizing a game: the *strategic form* and the *extensive form*. The strategic form makes use of the threefold structure we have already encountered: the list of players N, the list of strategies S_i available to each of them, and the payoffs $\pi_i(s)$ delivered to each of them corresponding to every admissible outcome s. Using these elements we may build up the matrix of a game. Matrix 2.3 gives a generic example of a 2×2 game, i.e., one in which there are two players, 1 and 2, each endowed with two strategies, (s_{i1}, s_{i2}).

Conventionally, in every cell of the matrix, I indicate first the payoff of the row player (agent 1), then that of the column player (agent 2). The extensive form representation is also known as the game tree, and dates back to Kuhn (1953).[5] Figure 2.1 portrays the tree of the game represented in strategic form in Matrix 2.3. In fact, it is one specific tree out of the possible representations of the same game in extensive form.

From top to bottom, the game tree departs from an *initial node*, called the *root* of the game, where player 1 picks either s_{11} or s_{12}, represented as separate branches leading to the *intermediate nodes* where player 2 has to decide what to choose, either strategy s_{21} or strategy s_{22}. The presence of a connection between the two intermediate nodes pertaining to agent 2 conventionally reveals that the game takes place under imperfect information. This amounts to saying that, when choosing a strategy, player 2 has no idea of the node in which player 1 has driven the game, because moves are simultaneous. If the intermediate nodes were disconnected, the game is one of perfect information, and player 2 knows which nodes he/she is in. That is, player 2 has observed the choice made by player 1 at the root, as the game is being played sequentially. In such a case, each of the intermediate nodes is called a *singleton*. At the base of the tree, there should appear a set of *terminal nodes* (in this case, four), where payoffs $\pi_i(s)$ pop up and are delivered to players corresponding to every admissible outcome. Usually, the explicit representation of terminal nodes is omitted, as in this case. However, terminal nodes are indeed important insofar as they identify every possible outcome of the game, as cells do in the matrix form. For instance, the first terminal

Matrix 2.3

		2	
		s_{21}	s_{22}
1	s_{11}	$\pi_1(s_{11}, s_{21}); \pi_2(s_{11}, s_{21})$	$\pi_1(s_{11}, s_{22}); \pi_2(s_{11}, s_{22})$
	s_{12}	$\pi_1(s_{12}, s_{21}); \pi_2(s_{12}, s_{21})$	$\pi_1(s_{12}, s_{22}); \pi_2(s_{12}, s_{22})$

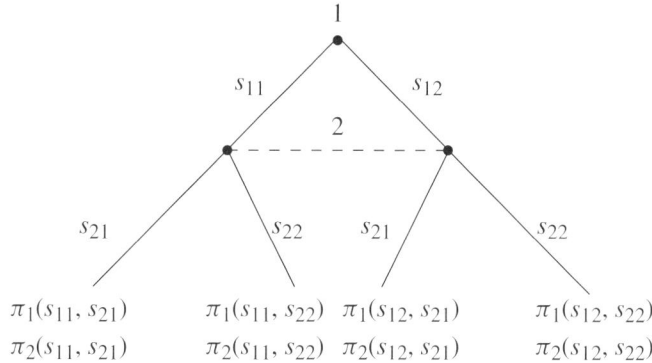

Figure 2.1 Extensive form of a 2×2 game with imperfect information

node on the left-hand side at the bottom of the tree identifies the outcome (s_{11}, s_{21}), with player 1 choosing to go left at the root and player 2 going left at the intermediate node. For later reference, it is the case of specifying that, when looking at extensive form games, the behaviour of players (that is, the list of strategies they play) is referred to as a *strategy profile*.

If the intermediate nodes are singletons, the subset of the game tree departing from each singleton constitutes a *proper subgame* of the entire game. A proper subgame must depart from a singleton differing from the root, and contains everything from that singleton all the way down to the terminal nodes that can be reached from that singleton itself. Of course, each game has an *improper subgame* that coincides with the entire game departing from the initial node, the latter obviously being a singleton by definition. In the example represented in Figure 2.1 this is indeed the case, as intermediate nodes are connected. Note that the imperfect information game illustrated in Matrix 2.3 and Figure 2.1 admits an alternative and fully equivalent extensive form representation that obtains simply by switching the relative positions of players, putting 2 at the root and 1 at the intermediate nodes. This holds true precisely because of simultaneous moves, whereby it doesn't really matter where a player appears along the game tree.

Conversely, under perfect information – or, sequential moves – the relative positions of players along the tree makes a lot of difference, as changing the order of moves entails looking at entirely different games, generating potentially diverging stories. The alternative games portrayed in Figures 2.2 and 2.3 are indeed two different games.

In Figure 2.2, player 2 has perfect information as he/she observes the strategy chosen by player 1 at the root. In Figure 2.3, the opposite holds, with player 2 moving first and player 1 endowed with perfect information corresponding to disconnected intermediate nodes. Observe that, conventionally, the payoffs are always listed in the same order (starting with player 1's) at each terminal node, irrespective of the order of moves.

Let's sum up what we have seen so far. A game of which we know the strategic form can also be outlined graphically as a tree, but the same matrix can generate several different (and not necessarily equivalent) trees, depending on the nature of the information players benefit from (or, whether moves are simultaneous or sequential). If information is perfect, then the player moving second observes the behaviour of the first mover and can choose a strategy accordingly. To anticipate some of the aspects that I will deal with when illustrating specific

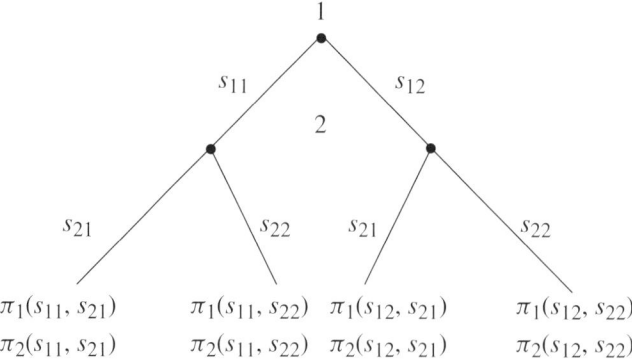

Figure 2.2 Extensive form of a 2 × 2 game with perfect information for player 2

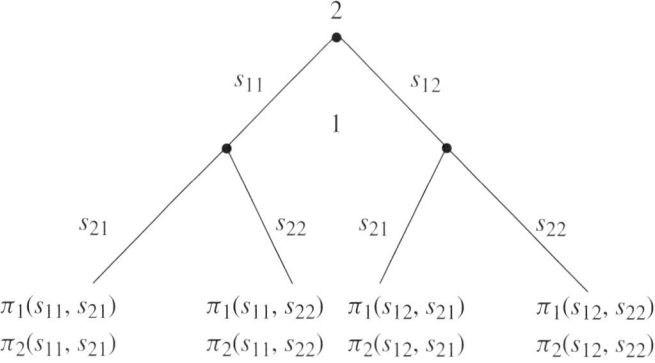

Figure 2.3 Extensive form of a 2 × 2 game with perfect information for player 1

games in extensive form in the remainder of the book, it is worth noting that the presence of perfect information (or sequential play) has two possible implications. The first is that, by moving first, the player located at the root may drive the development of the game in the desired direction, anticipating the reaction of the second mover. This is usually referred to as the *first-mover advantage*. The second implication is that playing second may in itself be not so bad, as it allows the second player to observe the rival's move and then decide what to do, possibly to the first player's disadvantage. As we shall verify throughout the following chapters, both possibilities may alternatively materialize depending on the specific features of the game.

Recall that information is complete if players know the structure of the game. In terms of its extensive form representation, this is equivalent to saying that, if the details of the game tree are fully known to players, then they play under complete information.

Now note that the extensive form conveys an amount of information about the nature of the game at hand definitely larger than the strategic (or matrix) form. Why? Because the tree signals whether moves are simultaneous or sequential, something that one is completely unable to infer on the basis of the strategic form, without being explicitly told so.

This, however, comes at a price. While the correct definition of a game in strategic form requires knowing the list of players, their respective strategy spaces and payoffs, the definition of the same game in extensive form requires knowledge of the following:

1   the entire game tree, starting from the root and ending up with terminal nodes, specifying the possible routes players can take all along the tree;[6]
2   the list of players;
3   the identity of the player that is required to move at every node;
4   the list of strategies available to each player corresponding to the node at which he/she is expected to move;
5   the list of payoffs accruing to players at each terminal node; and
6   the exact nature of information characterizing every intermediate node – that is, one must be able to tell whether any intermediate node is a singleton or not, and therefore also whether the game exhibits any proper subgames or not.

Having said all this, we are now in a position to proceed to the next issue, which is how to solve a game: the equilibrium concepts. The exposition will be confined to non-cooperative games.

Further reading

The material contained in this chapter can be further investigated by consulting several text-books in game theory, at different levels of formal complexity. At an intermediate level, see Gibbons (1992), Dutta (1999) and Osborne (2003). At a slightly higher level, see Montet and Serra (2003) and Rasmusen (2006). Finally, for classical reference books but at a somewhat advanced level, see Fudenberg and Tirole (1991), Binmore (1992), Myerson (1991) and Osborne and Rubinstein (1994).

3 Solving a game

The ball I threw while playing in the park
Has not yet reached the ground

Dylan Thomas,
Should lanterns shine

Being acquainted with the notions of game structure and information, we are now in a position to make a crucial step, which consists in identifying the solution concepts for non-cooperative games. These are usually defined as *equilibrium concepts*, or simply *equilibria*. What do we mean here (i.e., in the social sciences) by equilibrium? Our concept of equilibrium indeed reflects the corresponding concept that dominates physics. Trivially, any given system (no matter whether it is physical, biological, economic or social) is in a state of equilibrium if, in the absence of perturbations, it remains in its current state. With this in mind, of course, the investigation of equilibria in economic and social systems requires the analysis of the behaviour of agents driven by their respective incentives. What do people, parties or countries want to attain when they behave in a specific manner? The answer to this question, as we know from the previous chapter, is given by the payoff function assigned to each of them. The remaining issue, then, is to figure out the criterion informing their behaviour. This criterion is *individual rationality*. In game theory (or better, in the social sciences across the board), an outcome is individually rational if it gives each agent at least as much as his/her *security level*, the latter being the 'maximin' payoff – the maximum of all minimum payoffs – that this player can guarantee to him/herself irrespective of what the other agents do. But this is usually not enough. That is to say, what game theory assumes is that each player will rationally choose a strategy in order to pursue the *maximization* of his/her payoff, being aware of the relevant solution concept for the game being played, and the structure of the game itself. This leads to another crucial assumption, that of *common knowledge* across agents. This establishes that each player must be fully aware of the rules of the game, the nature of information in the game, and the game structure (including thus also the objectives of all other players); but this is not enough, as each player must know that every other player knows the same things, as well as that each player knows that everybody else knows, and so on recursively. Should the common knowledge assumption fail to apply consistently, then the game would not be correctly specified and the resulting analysis would yield completely unreliable predictions.[1]

Having said that, we may now take a full immersion into the most valuable and commonly used equilibrium concepts, starting with the oldest one, coined by John von Neumann.

3.1 The maximin (or minimax) equilibrium

The maximin (or minimax) equilibrium dates back to von Neumann (1928), but has been widely adopted as a solution concept after the publication of the book by von Neumann and Morgenstern (1944). This instrument is applicable to constant-sum games – while, as we shall see in the remainder, it is not applicable to variable-sum ones.

How does it work? The maximin version of the solution concept can be laid out as follows. Suppose we are looking at a two-player constant-sum game (of course, there may be more than two players, but this is the easiest way of making ourselves acquainted with the first solution concept we are meeting in the book). Each of them receives the following instructions: 'Given that this is a strictly competitive game, you have to choose the best strategy for yourself, in order to maximize the size of your slice of the pie you see in front of you, being aware that your rival is going to behave in a completely analogous way.' Or, equivalently: 'Given that this is a strictly competitive game, you have to choose the best strategy for yourself, in order to maximize the size of your slice of the pie you see in front of you, being aware that your rival is going to choose his/her strategy so as to minimize the size of your slice.' That is, given the constant-sum nature of the game, each player perceives the rival's behaviour as not only non-cooperative but indeed *aggressive*. Hence, each player's maximin strategy is chosen as a sort of insurance against the harmful consequences of the rival's attitude.

A bit more formally, the maximin strategy for player 1 guarantees to the latter the attainment of the maximum among the minimum payoffs generated by all admissible strategies available to 1 in the game at hand, given that player 2 is behaving in such a way so as to minimize 1's maximum payoff (whereby one may adopt the alternative label, minimax). That is, player 1 has to solve the following problem:

$$\max_{s_1} \min_{s_2} \pi_1(s_1, s_2) \tag{3.1}$$

while 2 is solving this:

$$\min_{s_2} \max_{s_1} \pi_1(s_1, s_2) \tag{3.2}$$

This, of course, is what 1 thinks that 2 is trying to do, while from the standpoint of 2 the problem is in fact to find the strategy that solves the problem $\max_{s_2} \min_{s_1} \pi_2(s_1, s_2)$, being convinced that 1 is trying to solve $\min_{s_1} \max_{s_2} \pi_2(s_1, s_2)$. Intuitively, a constant-sum game looks pretty much like a hall of mirrors. To clarify the operational meaning of von Neumann's equilibrium concept, we may resort to a simple example, described by Matrix 3.1.

The sum of the payoffs (the size of the pie to be split between players) is K. Note that, in view of the fact that we are looking at a constant-sum game, it would suffice to indicate player 1's payoff in each cell, as the payoff accruing to player 2 is just residually defined

Matrix 3.1 A constant-sum game

		2	
		s_{21}	s_{22}
1	s_{11}	3; $K-3$	5; $K-5$
	s_{12}	2; $K-2$	6; $K-6$

as the complement to K. Now we may identify ourselves with the row player, agent 1, and proceed to identify the minimum payoff for 1 along each of the two rows. This yields 3 along the top row (if 1 plays strategy s_{11}) and 2 along the bottom row (if he/she plays s_{12}). Following the maximin instructions, 1 should select strategy s_{11}, *expecting* to receive a payoff equal to 3. Now we can mimic the behaviour of player 2, who plays aggressively against 1 in order to minimize the latter's maximum payoff. By selecting strategy s_{21} (i.e., the first column), 2 spots 3 as the maximum payoff allowed to 1; instead, if 2 chooses strategy s_{22} (the right column), the maximum payoff accruing to 1 is 6. The minimum between the two maxima is 3, and therefore 2 should select s_{21} *expecting* the rival to get a payoff equal to 3. Thus, it appears that players' *ex ante* expectations are indeed *ex post* correct and reciprocally compatible.

Now take the opposite perspective, where player 1 is choosing the minimax strategy against player 2, while the latter is looking for the maximin strategy as an insurance: the maxima for player 2 along the two rows are, respectively, $K - 3$ (first row, strategy s_{11}) and $K - 2$ (second row, strategy s_{12}). Accordingly, player 1 should select strategy s_{11} in order to allow player 2 to get no more than $K - 3$. Along the columns, the minima for player 2 are $K - 3$ (first column, strategy s_{21}) and $K - 6$ (second column, strategy s_{22}). Consequently, 2 chooses s_{21} expecting to get $K - 3$. This is precisely what happens, and coincides with the outcome of the game if considered from the previous angle.

As a useful exercise, you may solve the game assuming that both players are following the same logic, either minimax or maximin, alternatively. This will allow you to quickly verify that the outcome is always the same irrespective of the combination of minimax and maximin behaviour attributed to the two players. Therefore, the outcome identified by the strategy pair (s_{11}, s_{21}) is the (unique) maximin or minimax equilibrium of this game.

If applied to a variable-sum game (which is not strictly competitive), von Neumann's equilibrium concept is bound to produce unacceptable results, as can be seen by inspecting the game in Matrix 3.2.

Here, maximin payoff is 3 for both players; as a consequence, 1 should play s_{12} while 2 should select s_{22}. By doing so, each of them would receive a payoff equal to 10. Of course 10 is higher than 3, and therefore players should be happy about the outcome generated by the maximin criterion.[2] However, a closer look reveals a number of issues putting into question the adoption of this equilibrium concept to solve a game like this. To begin with, the payoff generated by strategies (s_{12}, s_{22}) is not the maximin payoff of either player; consequently, the *ex ante* expectations of both players are not confirmed *ex post*. Moreover, how about the outcome (s_{11}, s_{21}), generating the highest payoffs appearing in the whole matrix? This is the Pareto-efficient (or Pareto-optimal)[3] outcome of the game – bluntly speaking, its happiest possible ending. Nonetheless, the criterion coined by von Neumann does not point to that: minimax and maximin behaviour actually leads players to dwell upon (s_{12}, s_{21}) and (s_{11}, s_{22}) because the corresponding cells contain their respective maximin payoffs, as well as (s_{12}, s_{22}) because this is the outcome generated by the maximin instruction, while (s_{11}, s_{21}) remains altogether outside of their conjectures. Such was the state of the art in

Matrix 3.2 A variable-sum game

		s_{21}	s_{22}
		2	
1	s_{11}	20; 20	2; 3
	s_{12}	3; 2	10; 10

Matrix 3.3

		2	
		s_{21}	s_{22}
1	s_{11}	20; 20	2; 3
	s_{12}	3; 11	10; 10

1944. Shortly after, a crucial advancement was made, in the form of what we now call the Nash equilibrium (Nash, 1950a, 1951).

3.1.1 *The Nash equilibrium*

Before looking at the Nash equilibrium concept, a preliminary step is in order. Observe again the game in Matrix 3.2, and assume that information is complete and symmetric (for the moment, it is immaterial whether information is perfect or imperfect, but let's say it is imperfect). Then, suppose player i formulates the following question: 'Given my rival's behaviour, what is my best strategy?' Or, equivalently: 'What shall I do to maximize my payoff, taking my opponent's strategy as given?' This question has two different answers, depending upon the choice made by player j. If the latter selects s_{j1}, then the optimal strategy is s_{i1} (because $20 > 3$); otherwise it is s_{i2} (because $10 > 2$). This mechanism generates what is labelled as player i's *best reply* (to the rival's behaviour). In the example related to Matrix 3.2, the best reply of either player changes depending on the strategy chosen by the opponent. This is not always the case – at least, not for all the players involved in a given game – as can be easily seen by looking at Matrix 3.3, where player 1's payoffs are the same as in Matrix 3.2, while player 2's payoff corresponding to the outcome (s_{12}, s_{21}) has been appropriately modified.

Here, the best reply of the row player takes the form of the following set of instructions: 'you should reply to s_{21} (respectively, s_{22}) by playing s_{11} (respectively, s_{12})', exactly the same as in the previous game. However, this is not true for the column player, as he/she realizes that s_{21} generates higher payoffs (i.e., is better) than s_{22}, *no matter what the row player is doing*. Accordingly, the best reply of 2 to 1 is to play s_{21}, relying on a sort of automatic pilot. Of course, in view of the assumptions of complete and symmetric information and common knowledge, player 1 is able to reconstruct *a priori* this feature of the game and therefore will expect player 2 to select s_{21}. Accordingly, 1 may confidently play s_{11}.

Now, keeping this discussion in the back of our minds, we may proceed to spell out the Nash equilibrium concept. There are several equivalent ways of defining it. The first is:

- **Nash equilibrium (version I)** Consider a non-cooperative game characterized by imperfect, complete and symmetric information, with a set N of players. An admissible outcome is a Nash equilibrium provided that no player has unilateral *ex post* regrets.

Let's have a closer look at the above definition. It says that an admissible strategy combination qualifies as a Nash equilibrium if and only if there exists no incentive for any of the players to deviate unilaterally from that outcome, taking the opponents' strategies as given and unmodifiable. This view of the Nash equilibrium requires that, if hypothetically each agent were asked in isolation after having played whether he/she would like to renege and change unilaterally the strategy just chosen, under the assumptions that the others stay put, the answer would systematically be negative for each player involved.

The second definition is the following:

- **Nash equilibrium (version II)** Consider a non-cooperative game characterized by imperfect, complete and symmetric information, with a set N of players. An admissible outcome is a Nash equilibrium if the strategy of each player qualifies as his/her best reply to the strategies selected by all the rivals.

This captures the idea that each player has to choose the strategy that maximizes his/her payoffs, being aware that all players are following the very same logic. Indeed, this requirement produces the same outcome that we would observe by imposing the absence of *ex post* regrets.

The formal definition that summarizes both the above intuitive versions is as follows.

DEFINITION 3.1 Given a non-cooperative game $G \equiv \langle N, S_i, \pi_i(s) \rangle$ with imperfect, complete and symmetric information, the outcome $\hat{s} \equiv (\hat{s}_1, \hat{s}_2, \ldots, \hat{s}_N)$, with $\hat{s}_i \in S_i$ for all $i \in N$, is a Nash equilibrium if and only if

$$\pi_i(\hat{s}_i, \hat{s}_{-i}) \geq \pi_i(s_i, \hat{s}_{-i}) \quad \text{for all } s_i \neq \hat{s}_i, \ s_i \in S_i$$

and this holds for all $i \in N$.

This amounts to saying that $\hat{s}$ is a Nash equilibrium if, *ex post*, no player i has a unilateral incentive to deviate from $\hat{s}$ unilaterally, given that the other players' behaviour $\hat{s}_{-i} \equiv (\hat{s}_1, \hat{s}_2, \ldots, \hat{s}_{i-1}, \hat{s}_{i+1}, \ldots, \hat{s}_N)$. This holds precisely because $\hat{s}_i$ is i's best reply to $\hat{s}_{-i}$.

Observe now the following simple statement contained in Nash's theorem.[4]

THEOREM 3.1 (NASH, 1951) *Every finite game $G \equiv \langle N, S_i, \pi_i(s) \rangle$ produces at least one Nash equilibrium in mixed strategies.*

It is worth noting that neither the definition of the equilibrium concept nor the existence theorem mention whether the game at hand is a variable-sum one or not. Indeed, this aspect is irrelevant, as the Nash equilibrium can be thought of as a generalization of von Neumann's minimax equilibrium, and coincides with the latter in constant-sum games.

To see this, reconsider the game in Matrix 3.1, from the standpoint of each player in turn. If player 2 selects s_{21}, 1's best reply is to play s_{11}. In such a case – that is, if 1 plays s_{11} – the optimal behaviour of 2 consists indeed in choosing s_{21}. Accordingly, the outcome (s_{11}, s_{21}), identifying the maximin (or minimax) equilibrium, is also a Nash equilibrium in pure strategies. At this point it is also easy to check that it is the *unique* pure-strategy Nash equilibrium of Matrix 3.1.

Now let's go back to Matrix 3.2, in relation to which we know that the minimax criterion is not applicable. Using the Nash equilibrium concept instead, this game is easily solved to yield two Nash equilibria in pure strategies, identified by the outcomes (s_{11}, s_{21}) and (s_{12}, s_{22}).

This result casts a shadow upon the Nash equilibrium concept: a finite variable-sum game may indeed produce a plethora of equilibria in pure strategies, leaving players (as well as external observers, like us) in what we might label at first sight as the *fog of strategic uncertainty* concerning the real unravelling of the game. What shall we expect to arise as the outcome of the game? One of the (many) Nash equilibria, or some other outcome generated by a *mistake*, something that *a priori* is not a Nash equilibrium? This problem, which is known in the literature on game theory as the issue of the multiplicity of equilibria, has generated a lively stream of research concerning the so-called *refinements* of the Nash equilibrium. These are more sophisticated (or, more demanding) solution concepts to be used so as to perform a selection among the Nash equilibria characterizing a given game, ideally isolating one Nash equilibrium that, for some acceptable reasons, looks more solid or more credible than the others.

3.2 Refinements of the Nash equilibrium

Carrying out an exhaustive overview of equilibrium refinements is probably an overwhelming and ultimately hopeless task, as their list is growing steadily on an almost daily basis. Moreover, this objective would not be in accord with the aims of the present volume. For these reasons, I will confine my attention (and yours) to the few refinements that are more intuitive than others and also most commonly used:

- equilibrium in dominant strategies;
- focal point equilibrium;
- subgame perfect equilibrium; and
- risk dominance (in a section of its own).

3.2.1 Equilibrium in dominant strategies

This refinement is based on an idea that we have already encountered at the outset of this chapter, that is, the possibility that a player's best reply be invariant to the rivals' strategies. If this sort of automatic pilot does work, the resulting best reply qualifies as that player's dominant strategy. The rigorous definition is as follows.

DEFINITION 3.2 Strategy $\tilde{s}_i \in S_i$ is (at least weakly) dominant for player i if it maximizes his/her payoff $\pi_i(\tilde{s}_i, s_{-i})$ irrespective of the opponents' behaviour s_{-i}, that is, if and only if

$$\pi_i(\tilde{s}_i, s_{-i}) \geq \pi_i(s_i, s_{-i}) \quad \text{for all } s_i \neq \tilde{s}_i, \ s_i \in S_i$$

If the above inequality holds strictly, then $\tilde{s}_i$ is *strictly dominant*; otherwise, if it holds as an equality for at least some admissible s_{-i}, then it is *weakly dominant*.

In agreement with Definition 3.2, if all players do have a dominant strategy (or at least one that is weakly so) and adopt it, the outcome identified by the combination of (at least weakly) dominant strategies will qualify as an equilibrium in dominant strategies, as follows.

DEFINITION 3.3 Given a game $G \equiv \langle N, S_i, \pi_i(s) \rangle$, the outcome $\tilde{s} \equiv (\tilde{s}_1, \tilde{s}_2, \ldots, \tilde{s}_N)$, with $\tilde{s}_i \in S_i$ for all $i \in N$, is an equilibrium in (at least weakly) dominant strategies if and only if

$$\pi_i(\tilde{s}_i, s_{-i}) \geq \pi_i(s_i, s_{-i}) \quad \text{for all } i \in N$$

This concept deserves a few comments. In particular, it is worth dwelling upon the difference between the condition that must be satisfied for an outcome to be a Nash equilibrium and the one contained in Definitions 3.2 and 3.3 in order for a dominant strategy equilibrium to arise. While the former requires a player to identify the best reply to every possible strategy chosen by any rival – whereby no *ex post* regrets may exist – the latter poses a much stronger requirement, which consists in finding a strategy yielding a payoff systematically at least as high as that for any other strategies the same player could adopt, no matter what the opponents do. Relatedly, as we shall see below, the existence of an at least weakly dominant strategy for a player does not necessarily entail that this player's best reply will be unique. To put it differently, this indeed entails that we commonly observe games with several Nash equilibria in pure strategies, generated by the crossing of best replies, and, if we're lucky, one of these equilibria will emerge as an equilibrium in at least weakly dominant strategies. Equivalently, we may say that *while every dominant strategy equilibrium is also a Nash equilibrium, the opposite is not true*. In the light of these considerations, it should be clear that the dominance criterion is much more demanding than the no *ex post* regrets criterion embodied in the Nash equilibrium concept.

Having said that, it is now time to put aside linguistic tricks and proceed to see how dominance works. To this end, I will illustrate a very famous game, the so-called *prisoners' dilemma*. This is represented in Matrix 3.4.

The situation described in this game is usually spelled out as the following traditional story. The players are two criminals who are suspected of having committed a crime (say, a bank robbery by night) together. The police caught them, and brought them in front of the district attorney. However, evidence is lacking, and therefore the district attorney may only rely upon a full confession on the part of at least one of the prisoners. Therefore, the two racketeers are locked into two separate cells to prevent any communication between them, and the attorney interrogates them in turn, hoping to obtain a confession.

Formally, this picture yields a non-cooperative variable-sum game, with imperfect, complete and symmetric information.[5] Prisoners have two pure strategies, either *to confess* (*c*) or *not to confess* (*nc*). The payoffs appearing in the matrix represent the time spent in jail depending on the outcome. Hence, they are all negative, with a relevant exception concerning what happens if one prisoner confesses while the other doesn't.

Clearly, both prisoners are aware that they'd better not say anything, as in such a case (i.e., corresponding to (*nc, nc*)) they would be readily released. However, they cannot communicate and therefore cannot rely on credible agreements between them, and the key of the game lies in the interpretation of the asymmetric outcomes along the secondary diagonal of the matrix, where one is confessing while the other is not. The story usually tailored to fit these outcomes is that the attorney offers to each prisoner the possibility of leaving the police station immediately as a free man (provided the other is keeping his/her mouth shut) after signing a full confession recognizing the responsibility of both. This makes strategy *nc* strictly dominant over strategy *c*, as each prisoner hopes to attribute the full punishment to the other and go back to the pub to play pool and drink beer. Unfortunately for the prisoners, this mechanism selects outcome (*c, c*) at the equilibrium in dominant strategies of the game, which is also its unique Nash equilibrium in pure strategies, as can be easily ascertained.

A few additional observations will help shed some further light on the striking relevance of a game like the prisoners' dilemma.

Matrix 3.4 The prisoners'
dilemma

		2	
		nc	*c*
1	*nc*	−1; −1	−10; 0
	c	0; −10	−5; −5

- Its formal structure is very common in real-world situations that are surely familiar to all of you. To illustrate this point, consider first that the driving force behind the disastrous equilibrium outcome is called the *free-riding incentive*. Why? Because each prisoner confesses for the very same reason many people take the bus without paying tickets every day. It is also the same reason why anyone may decide to keep the air conditioner in the office on around the clock, seven days a week, even if this clearly contributes to global warming. At this point, it should be clear that the games in Matrices 2.1 and 2.2 are prisoners' dilemma.
- Relatedly, the equilibrium is Pareto-inefficient. The prisoners' dilemma is the prototypical structure to be used to illustrate that selfishness (or, more politely, rational non-cooperative maximization of individual objectives) is bound to produce socially suboptimal outcomes. Free-riding incentives and Pareto inefficiency are intimately connected. The idea that binding agreements are ruled out is in some sense implicit in the non-cooperative nature of the game, preventing the players from generating (nc, nc) as the equilibrium. The possibility that a prisoners' dilemma may indeed yield Pareto-efficient equilibria, and the conditions that must hold for this to happen, will be investigated in Chapter 5.

The second game that we shall take into consideration to become acquainted with the dominance criterion is in Matrix 3.5, where the players' strategies are $(t = top, b = bottom)$ for 1 and $(l = left, r = right)$ for 2.

This game produces two Nash equilibria in pure strategies, (t, l) and (b, r). This is easily checked: if 1 plays t, the best reply of 2 is to play l, and conversely. Suppose now 2 chooses r. In such a case, 1's best reply is b; given the strategy played by 1, 2 has no incentive to deviate towards l, as this yields exactly the same payoff (remember that a unilateral deviation must bring about a strictly positive gain for a player to deviate). Hence, both outcomes along the main diagonal of the matrix are pure-strategy Nash equilibria. At first sight, this poses a question as to the actual outcome of the game, as, say, player 1 selects t expecting to generate the equilibrium (t, l) while the rival chooses r expecting the equilibrium (b, r) to obtain. Observe though that, while player 1 has no dominant strategy, there is a weakly dominant strategy for player 2, which is *left*, since $2 > 0$ and $3 = 3$. On this basis, player 1 may expect 2 indeed to select *left*, in which case the right column of Matrix 3.5 becomes immaterial. This generates a reduced form of the game as in Matrix 3.6, where 1 identifies top as his/her strictly dominant strategy, since $6 > 3$.

Matrix 3.5

		2	
		l	*r*
1	*t*	6; 2	0; 0
	b	3; 3	3; 3

Matrix 3.6

		2
		l
1	*t*	6; 2
	b	3; 3

This leaves us with a further reduced form game, consisting of the single outcome (t, l) that qualifies as a Nash equilibrium in (weakly) dominant strategies, attained by *iterated deletion* of (weakly) dominated strategies. It is worth stressing that the assumptions of common knowledge and complete information allow both players to reconstruct this line of reasoning in advance (that is, before actually playing the game). The expected result is that the column player will indeed disregard the right column as it represents a weakly dominated strategy, allowing then the iteration taking place in the minds of both players to trigger an equilibrium selection process yielding (t, l) as a credible outcome. The latter is more credible or more solid than (b, r) precisely because it doesn't involve the adoption of dominated strategies (or at least weakly so). Note, however, that (b, r) *remains a Nash equilibrium* and therefore cannot be dismissed as irrelevant. Several good reasons for not disregarding it altogether will become clear in the remainder of this chapter.

To further illustrate how iteration works, look at the non-cooperative game in Matrix 3.7, where information is again complete, symmetric and imperfect. This game has two Nash equilibria in pure strategies, (t, l) and (m, r). From the standpoint of player 2, strategy r weakly dominates strategy l, and the equilibrium (m, r) is the intersection of weakly dominant strategies, attained by iteration. Indeed, player 1 will dismiss b as it is strictly dominated by both t and m. Having eliminated the left column, what remains of the matrix reveals that m dominates t along the right column.

However, the above discussion does not imply that dominance is a definitive solution to the issue of equilibrium selection in matrix games with imperfect information. Several famous examples will clarify this aspect. The first is depicted in Matrix 3.8, and it is traditionally known as the *battle of the sexes*. The two players, he and she, are dating each other. To keep things as simple as possible, let's say that they may go either to the movies (m) or to the stadium (s). She prefers the first option, while he prefers the second; yet, both prefer to go to the same place together rather than to different ones alone. These reasonable assumptions underlie the sequence of payoffs.[6] Both outcomes along the main diagonal, (m, m) and (s, s), are pure-strategy Nash equilibria. Yet, dominance doesn't help us here, as neither player exhibits a dominant strategy (or at least weakly so).

Another game with similar properties is the so-called *chicken game* that appears in Matrix 3.9. The players are, say, Kurt Russell (K) and Sylvester Stallone (S) driving their two magnificent trucks. They find each other at the opposite ends of a very narrow bridge without road lights. Both of them are endowed with a very strong personality and care about

Matrix 3.7

		2	
		l	*r*
	t	2; 8	2; 8
1	*m*	0; 1	4; 2
	b	−1; 3	1; 7

Matrix 3.8 The battle of the sexes

		He	
		m	*s*
She	*m*	2; 1	0; 0
	s	0; 0	1; 2

Matrix 3.9 The chicken
game

		c	nc
		S	
K	c	0; 0	2; 1
	nc	1; 2	0; 0

the consideration of the community of truck drivers. Therefore, each would like to be the first to cross the bridge in order not to be marked as a chicken (i.e., a coward) by his fellow truck drivers. The resulting pure strategies are to cross (*c*) or not to cross (*nc*).[7]

Clearly, every asymmetric outcome along the secondary diagonal identifies a Nash equilibrium in pure strategies, and we cannot resort to dominance as an equilibrium selection mechanism. Games like these are usually labelled as *coordination games*, as they describe situations generating a multiplicity of Nash equilibria that requires the adoption of some commonly accepted coordination devices: habits, traditions, rules, laws. In the case of Kurt and Sylvester, a simple road light will do. For the couple playing the battle of the sexes, a reasonable rule (and indeed one that is very often adopted, also according to my personal experience) consists in alternating *movie* and *stadium* on a weekly basis, depending on the calendar of the football championship, right?

The third example in the field of coordination games is the so-called *stag hunt game*. In some sense, this is a noble example, as it dates back to Rousseau.[8] A tribe of early *Homo sapiens* is setting off to hunt game (sorry for the pun!), consisting of either stag (*s*) or rabbit (*r*), for simplicity. Two hunters, 1 and 2, are in charge of feeding their people, knowing that: (i) each hunter can surely come back to the village every day with a rabbit; while (ii) it takes both of them to get the stag, because if only one hunter tries his luck after it, the stag runs away unscathed; and finally (iii) a stag is worth more than two rabbits. Say, the nutrition facts of our stylized prehistoric tale establish that you need ten rabbits to make up for a stag. This situation is represented in Matrix 3.10.

A quick look at the matrix suffices to see that there are two Nash equilibria in pure strategies, (*s*, *s*) and (*r*, *r*), and the game cannot be solved in dominant strategies. This is in fact the same structure as emerges from Matrix 3.3, whose essential feature is that the two Nash equilibria can be Pareto-ranked. The same applies here: the tribe would be better off if its hunters went for a stag, but Pareto-efficiency in itself does not ensure that they won't waste their time killing rabbits instead, then coming back to the village with comparatively little food. That is, the lack of relationship between Nash and Pareto prevents us, in general, from invoking efficiency as a criterion for equilibrium selection. Of course, there is a way out of the coordination issue posed by Matrix 3.10, coinciding with Rousseau's suggestion: the chief (or the council) of the tribe may solve the problem by adopting a rule establishing that hunters should forget about rabbits and concentrate on the stag (possibly with drastic consequences if they don't).

Matrix 3.10 The stag hunt
game

		s	r
		2	
1	s	5; 5	0; 1
	r	1; 0	1; 1

Matrix 3.11

	2	
	l	*r*
t	10; 25	1; 25
m	4; 10	9; 10
b	−3; 8	−7; 1

1

The games we have reviewed thus far belong to two categories, one collecting games where iterated dominance works and selects one out of several Nash equilibria, and the other collecting games where the dominance criterion simply doesn't apply (like the battle of the sexes or the stag hunt game). There are also other games lying in a sort of no-man's land, where the iterated deletion of at least weakly dominated strategies works – at least to some extent – but apparently yields weird results.

Consider the game represented in strategic form in Matrix 3.11. This is a 3×2 game yielding two pure-strategy Nash equilibria, (t, l) and (m, r). Can the iterated deletion of (at least weakly) dominated strategies help delete any rows or columns, so as to ultimately select a single equilibrium?

This turns out to be an ambiguous procedure in the present game. Consider first the column player, agent 2. Strategy l weakly dominates r, and therefore the latter can be deleted. In the residual 3×1 matrix, player 1 selects t as a strictly dominant strategy. Hence, the iterated deletion process seemingly points at (t, l) as the credible equilibrium.

This conclusion, however, is doomed to be contradicted if we start deleting dominated strategies taking the row player's perspective first. In such a case, we see that, given that strategies t and m strictly dominate strategy b, the latter can be deleted. Looking at the remaining 2×2 matrix, however, player 2 realizes to be just indifferent between l and r (and also player 1 is aware of that). Hence, they get stuck into a 2×2 matrix yielding the same two Pareto-rankable equilibria already characterizing the original larger matrix. The delicate issue here is that what we imagine (the process of deleting dominated strategies) does not take place in the game, but rather in the players' minds *beforehand*. Therefore, each of them has to conjecture about the fictitious order of deletion that the opponent is envisaging, and the two orders may not be reciprocally compatible.[9]

This brief overview of relatively simple games gives a rough idea of the problems encountered by players in trying to envisage the equilibrium outcome of a given game. But their life can be much worse than this, as many games are far more complex than the ones we have seen so far. The next refinement tackles precisely this aspect.

3.2.2 Focal point equilibrium

The notion of focal point equilibrium dates back to Schelling (1960). It can be understood in purely intuitive terms on the basis of the following informal definition. Suppose a game produces a large number of Nash equilibria (i.e., in general, strictly more than two) in pure strategies; if there is one emerging from the bunch of equilibria for any plausible reason because it is different from all the others, this equilibrium qualifies as the *natural solution* or focal point of the game. This is illustrated in Matrix 3.12.

This is a 'large' game, where each player may choose among n strategies. The resulting $n \times n$ matrix (of which only the 5×5 top-left portion is represented) is characterized by the fact that any strategy pair identifying a cell along the main diagonal is a Nash equilibrium.

Matrix 3.12 A 'large' coordination game

		s_{21}	s_{22}	s_{23}	s_{24}	s_{25}	$\ldots$
	s_{11}	100; 100	$-1; -1$	$-1; -1$	$-1; -1$	$-1; -1$	$-1; -1$
	s_{12}	$-1; -1$	100; 100	$-1; -1$	$-1; -1$	$-1; -1$	$-1; -1$
1	s_{13}	$-1; -1$	$-1; -1$	7; 7	$-1; -1$	$-1; -1$	$-1; -1$
	s_{14}	$-1; -1$	$-1; -1$	$-1; -1$	100; 100	$-1; -1$	$-1; -1$
	s_{15}	$-1; -1$	$-1; -1$	$-1; -1$	$-1; -1$	100; 100	$-1; -1$
	$\ldots$	$-1; -1$	$-1; -1$	$-1; -1$	$-1; -1$	$-1; -1$	100; 100

(The column header group is labelled **2**.)

The peculiar feature of the set of Nash equilibria is that all of them, except one (the pair (s_{13}, s_{23})) yield the same payoffs. This creates, in principle, a huge coordination problem to players acting under imperfect information, as the probability of making systematic mistakes is extremely high: if, say, 1 is considering the adoption of strategy s_{12} under the conjecture that 2 will play s_{22}, while the latter is conjecturing to play s_{25} (or any strategy other than s_{22}), the outcome will be outside the main diagonal.

Here Schelling's focal point concept comes into the picture to help players solve the coordination problem, by noting that (s_{13}, s_{23}) is *different* from any other Nash equilibrium, as if it were under a strong spotlight attracting the players' attention. Note that (s_{13}, s_{23}) qualifies as the focal point equilibrium of Matrix 3.12 simply because of its peculiarity, not because it is any better than any other Nash equilibrium – in fact, in this example it is much worse!

This argument reinforces the bottom line of the foregoing discussion on coordination games where Nash equilibria can be Pareto-ranked (as, e.g., the stag hunt game), namely, that there is no relationship between the criterion of individual rationality underlying the Nash equilibrium concept and that of Pareto efficiency, the latter being obviously palatable but far from offering an easy way out of the multiplicity problem.

So much for coordination games. No, something extremely relevant remains to be said about this class of games. We may further investigate them using another refinement of the Nash equilibrium concept, which has quickly become one of the main instruments of game theory, and has largely contributed to produce a wide range of useful applications in all areas of the social sciences.

3.2.3 *Subgame perfect equilibrium by backward induction*

> We shall not cease from exploration
> And the end of all our exploring
> Will be to arrive where we started
> And know the place for the first time
>
> T. S. Eliot,
> *Little Gidding (Four quartets)*

So far, we have examined only games in strategic form. Now, we enter the realm of games represented in extensive form, the so-called Kuhn tree. The concept of subgame perfect equilibrium (Selten, 1965, 1975) is indeed defined in relation to the extensive form of a game.

DEFINITION 3.4 Given a game $G \equiv \langle N, S_i, \pi_i(s) \rangle$, the outcome $\bar{s} \equiv (\bar{s}_1, \bar{s}_2, \ldots, \bar{s}_N)$, with $\bar{s}_i \in S_i$ for all $i \in N$, is a subgame perfect equilibrium by backward induction if it induces a Nash equilibrium in every proper subgame of the original game.

At first sight, this definition may sound somewhat obscure. A few terms appearing in the definition deserve some careful explanation. The first is *backward induction*. In practice, a subgame perfect equilibrium must be identified by using the backward induction method, i.e., proceeding from bottom to top along the extensive form of the game. Bluntly speaking, backward induction is the process of reasoning backwards in time, to determine a sequence of optimal actions. It proceeds by first considering the last time a decision has been taken and then choosing what to respond rationally to such a decision. Using this information, one can then determine what to do at the second-to-last time a decision has been taken, and so on and so forth. This process keeps going backwards until one has determined the best action at every point in time (along a game tree, at every node). In a game, the backward induction process starts from the terminal nodes to reach the root of the tree, examining the optimality of decisions taken at each and every intermediate node, no matter how many of these there are. The second element requiring our attention is the concept of *proper subgame*. As we already know from Chapter 2, a proper subgame is a subset of a game tree departing from a singleton, which in turn is an intermediate node unconnected with any other node at the same level. Having said that, it should be clear that the elective field of application of the subgame perfect equilibrium is the realm of games with perfect information, that is, those in which players move sequentially, each of them thus being able to correctly recollect the entire past history of the game up to the point where he/she is required to choose a strategy. Also, these are the games that lend themselves to a graphical representation in the form of Kuhn's trees, which of course helps to visualize the solution generated by backward induction.

To grasp the operational contents of the notion of subgame perfection, consider the whole set of admissible extensive forms generated by the game we have already encountered in strategic form in Matrix 3.5.

The first (Figure 3.1) describes the case where player 2 moves after player 1, and therefore enjoys perfect information. Note that the subgame starting from the right singleton is not pictured as choice of strategy *b* by player 1 makes the subsequent behaviour of player 2 irrelevant in that the payoff pair is surely (3, 3). Accordingly, focus your attention on the left subgame, which becomes relevant if player 1 chooses *t*. In this subgame, player 2 may get either 2 or 0: since 2 > 0, you may expect player 2 to select *l* as a best reply to *t*. With this in mind, you may now proceed backwards to examine player 1's conjectures at the root. Having in mind that (i) player 2 will respond to *t* by choosing *l*, while (ii) he/she will be

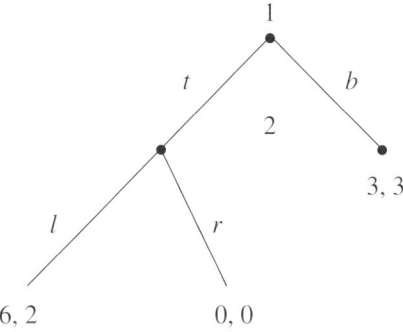

Figure 3.1 Perfect information for player 2

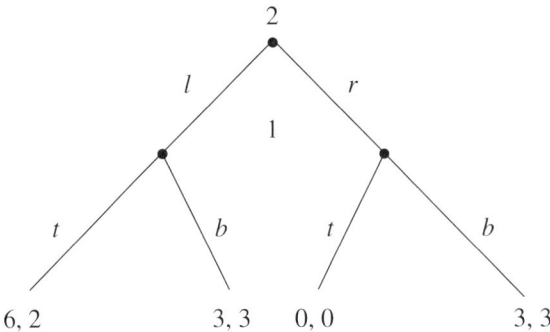

Figure 3.2 Perfect information for player 1

indifferent between *l* and *r* if player 1 chooses *b*, 1 is facing the alternative of getting either 6 by playing *t* or 3 by playing *b*. Since 6 > 3, player 1 will prefer to play *t*. Therefore, the subgame perfect equilibrium of the game where information is perfect for player 2 is (*t, l*). It is worth stressing that this coincides with the Nash equilibrium in (weakly) dominant strategies previously identified by the examination of the strategic form of the same game.

The alternative situation, where information is perfect for player 1 (because player 2 is at the root) is represented in Figure 3.2 (to avoid confusion, the order of payoffs is the same as in the original matrix and Figure 3.1). Here both subgames must be drawn explicitly, as player 1 is never indifferent between *t* and *b*. By backward induction, we see that the best reply to *l* is *t*, while the best reply to *r* is *b*. Hence, player 2 knows that he/she will get 2 by playing *l* and 3 by playing *r*. Accordingly, in this case the subgame perfect equilibrium will be (*r, b*), i.e., the Nash equilibrium that one would dismiss as involving the adoption of a weakly dominated strategy on the basis of the above analysis of Matrix 3.5.

This simple example shows that, in general, *the subgame perfect strategy profile is sensitive to the sequence of play*, and the fact that a Nash equilibrium involves the use of a (weakly) dominated strategy does not entail that there is no sequence of play under which that outcome can qualify as a subgame perfect equilibrium. This shows, additionally, that under sequential play there exists a *first-mover advantage* whereby the player entitled to move before the other(s) may drive the game in the direction of a specific Nash equilibrium among those generated by the game at hand.

These properties of perfect information games play a crucial role that will become apparent when we discuss applications to economics and politics in the next chapter. Here, we may further dwell upon these features of sequential play by re-examining some of the games we have already seen in the first part of this chapter. Before doing this, it is just the case of asking ourselves what happens if we try to use backward induction and subgame perfection in a game with imperfect information. To this end, have a look at the extensive form game appearing in Figure 3.3. This is the tree generated by Matrix 3.5 if players move simultaneously.[10] Since intermediate nodes are connected, the game possesses a unique improper subgame, coinciding with the tree itself. Hence, according to Definition 3.4, any Nash equilibrium of the original game appears subgame perfect here, which amounts to saying that subgame perfection is applicable but fails to operate the desired selection between the two

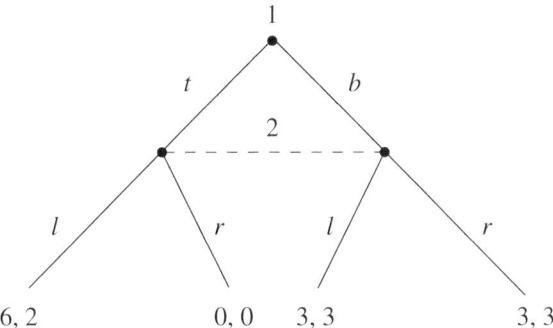

Figure 3.3 Imperfect information (simultaneous play)

Nash equilibria (or equivalently, it is no longer a refinement). This should clarify once and for all that subgame perfection is a useful selection tool if and only if the game is characterized by perfect information.

Now let's reconsider the battle of the sexes, and, ladies first, let's suppose that she is at the root and moves first. This extensive form game is shown in Figure 3.4 (again, the order of payoffs at every terminal node is the same as in Matrix 3.8).

Corresponding to each intermediate note, the best reply of the man is to adopt the same strategy adopted by her at the root, for obvious reasons. Therefore, the woman's perception of the game from the initial node induces her to move in the direction of the movie theatre. So, if she plays first, the subgame perfect equilibrium is (M, M). Of course, the opposite applies if the man moves first, in which case subgame perfection selects (S, S). Therefore, the subgame perfect equilibrium is indeed unique, for a given sequence of play.

An even more interesting (and relevant) result emerges by reconsidering the stag hunt game played under perfect information. Assume hunter 1 is at the root (that is to say, suppose he leaves the village before hunter 2). This case is portrayed in Figure 3.5.

Clearly, the presence of perfect information selects (s, s) as the unique subgame perfect information, because the hunter located at the intermediate nodes finds it optimal to choose

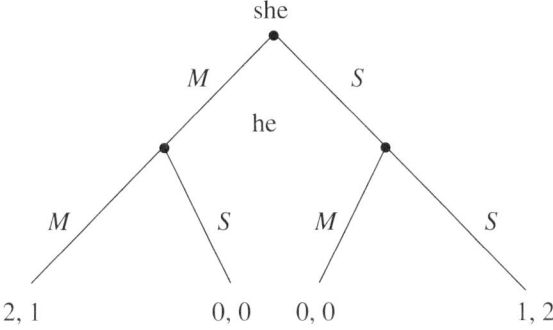

Figure 3.4 The battle of the sexes (ladies first)

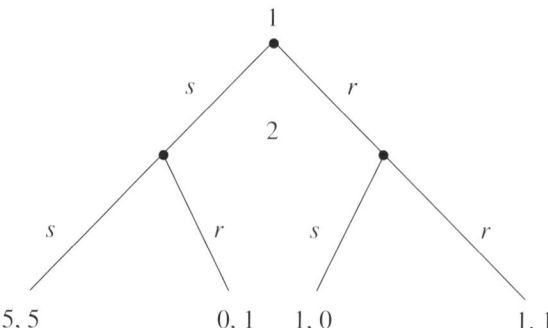

Figure 3.5 The stag hunt game under perfect information

(i) strategy *s* having observed the first mover playing *s* (this happens at the left singleton) or (ii) strategy *r* having observed the first mover playing *r* (this happens at the right singleton); hence, the hunter at the initial node selects *s* because $5 > 1$.

This is an intuitive example pointing in the direction of a general result, which I will state without further unnecessary and involved proof.

THEOREM 3.2 *In a game generating multiple Nash equilibria that can be Pareto-ranked, the adoption of sequential play (or equivalently, the presence of perfect information) leads to the selection of the Pareto-efficient equilibrium as the subgame perfect solution, irrespective of the order of moves followed by players.*[11]

This theorem conveys a very powerful message, namely, that increasing the amount of information available to players – in particular, making the game one of perfect information via sequential play – generates an equilibrium selection process resulting in individual rationality (i.e., selfishness) going hand in hand with social efficiency. This mechanism, however, operates only in the relative restricted class of games yielding multiple equilibria on which the order of preference is exactly the same across the population of players. This, unfortunately, immediately entails that sequential play is no medicine for the prisoners' dilemma, where the Nash equilibrium is *unique*.

There is more to it: sequential play, backward induction and subgame perfection may yield equilibria in pure strategies even in games where Nash equilibria under simultaneous play indeed fail to exist. This is an intriguing and a bit more advanced topic, to which I will come back when dealing with mixed-strategy equilibria, in the last section of this chapter.

3.3 Warnings

At this point, it is just fair – and wise – to pause a little and take time to reflect on a few critical aspects of what we've seen so far. In particular, I will draw your attention to two famous games that invite us to use game theory and take the indications yielded by the theory itself with extreme care, and a pinch of salt. The first of these games is aimed specifically at the backward induction mechanism, while the second has a more general flavour, as it tackles the issues of individual rationality and maximizing behaviour.

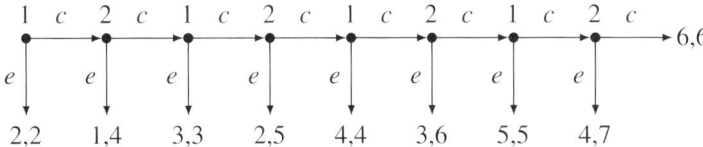

Figure 3.6 The centipede game

3.3.1   The centipede game

The so-called *centipede game* (Rosenthal, 1981) is a widely used example to highlight that backward induction and the associated subgame perfect equilibrium concept (as well as, for completeness, the iterated deletion of dominated strategies) may sometimes lead one to draw unreliable theoretical predictions concerning the outcome of games with perfect information. The centipede is an extensive form game with two players acting sequentially, each one for the same number of rounds, along a tree like the one illustrated in Figure 3.6.

At every node, a player may choose either to *exit* (e) or to *continue* (c). Agents act under perfect, symmetric and complete information. Reasoning by backward induction, starting from the last node, one should infer that player 2 will select e because $7 > 4$. One node back, player 1 rationally anticipates this behaviour on the opponent's part, and therefore must choose e because $5 > 4$. That is, at every node strategy e strictly dominates strategy c and this applies as well to the initial node, at which 1 should exit immediately. Therefore, the subgame perfect equilibrium requires the player moving at the root to exit, so that the entire tree will not materialize and the other agent will never play at all.

Very much like the ultimatum game, the centipede game is also designed to send out a warning message against the indiscriminate use of our instruments – or, it warns that one should not blindly trust the prediction of a theoretical set-up based on rationality and (purely selfish) maximization. Indeed, experimental results[12] indicate that the outcome qualifying as the unique subgame perfect equilibrium of the game is almost never confirmed in practice, irrespective of the population of players that is employed to carry out the experiment (students, chess players, civil servants, etc.). The actual distribution of observed outcome depends on the size of the stakes and the number of rounds, so that in general the game does unravel for a number of rounds before anyone decides to exit.

However, the way in which backward induction drives agents' behaviour in the centipede game can also be interpreted in another way – possibly much more intriguing than the aforementioned warning. If you go back for an instant to note 3 in Chapter 1, and read again von Neumann's statement as to the plausibility of a preventive nuclear strike against the USSR, you will certainly appreciate its genuine backward induction flavour and its close similarity to the theoretical (if not practical) solution of the centipede game.

3.3.2   The ultimatum game

The ultimatum game is a one-shot two-player non-cooperative game, with perfect, symmetric and complete information, where a player (labelled as player 1) is given a certain amount of money, say, 100 euro, and has to split this sum between himself and player 2. The latter can either accept or reject 1's proposal. If player 2 rejects the offer, neither player receives anything. If, instead, player 2 accepts it, the initial amount of money is split between them as proposed by player 1. Both players are fully aware of all this. Given that the minimum admissible amount that 1 can offer to 2 is one cent, this would indeed be the proposal that 1 should decide to do under the assumptions of individual rationality and maximizing behaviour: if one takes these assumptions at face value, player 2 should accept such an offer, since one cent is better than nothing.

Yet, experimental results prove that this is *almost never* the case. Indeed, the vast majority of people playing the role of player 1 in experiments carried out on the ultimatum game (with real money) propose their counterparts something between 30 per cent and 60 per cent of the initial sum.[13] Why? For many different but equally good reasons, like altruism, fairness, empathy and risk aversion. Player 2, facing the perspective of going home with a single cent, might well have fun in rejecting the proposal just for the sake of melting away the 99.99 euro that player 1 is so lavishly expecting to get. After all, one cent is more than nothing, but is also very close to nothing!

The experimental evidence generated through the ultimatum game is generally considered as strong evidence against the *Homo oeconomicus*, as seen through the magnifying lenses of rationality and maximization. My personal view is a little milder. As in the case of the centipede game, a warning is a warning, and is not necessarily valid across the board. Were it so, then you might as well stop reading the book at this point. Please do not, as the best is yet to come.

Before proceeding any further, it will be useful to draw some synthetic considerations on the toolkit we have reviewed thus far. To begin with, a finite non-cooperative variable-sum game may yield several Nash equilibria. Hence, the Nash equilibrium appears not to be such a demanding solution concept – indeed, the problem with it is that it works far too well, and too often. Consequently, researchers have designed other, and more sophisticated, solution concepts – like the equilibrium in dominant strategies and the subgame perfect equilibrium – aimed at operating a selection out of the *riches* of Nash equilibria, in order to dismiss all of them but one that qualifies as more credible or robust according to some acceptable criteria. Is it possible to assess and rank the strength and effectiveness of subgame perfection and dominance? To some extent yes, provided that each of the two solution concepts operates in its own elective domain (the dominance criterion is usually adopted to solve games in strategic form under imperfect information, while subgame perfection is a meaningful concept only in extensive form games with perfect information). This appraisal is carried out in the following theorem, of which – as usual, by now – I will not provide a formal proof.

THEOREM 3.3 *In a game $G \equiv \langle N, S_i, \pi_i(s) \rangle$ yielding multiple Nash equilibria in pure strategies, the following statements hold true.*

(i) *An equilibrium in (at least weakly) dominant strategies may fail to exist. If it exists, it is an element of the set of subgame perfect equilibria generated by every admissible sequence of moves.*

(ii) *The set of subgame perfect equilibria is a subset (sometimes, a proper one) of the set of Nash equilibria.*

We are now in a position to make a couple of additional steps and deal with probabilities and risk.

Matrix 3.13

$$
\begin{array}{c|cc}
 & & 2 \\
 & l & r \\
\hline
1 \quad t & 10;10 & 0;5 \\
b & 5;0 & 7;7 \\
\end{array}
$$

3.4 Risk dominance

What if a player is uncertain about the other players' strategies? This issue is particularly relevant in games yielding more than one Nash equilibrium in pure strategies, i.e., all games involving coordination issues. Consider a coordination game like the one illustrated in Matrix 3.13.

This game has two Nash equilibria in pure strategies, (t, l) and (b, r), the former being the Pareto-optimal outcome. However, as we already know, Pareto efficiency does not necessarily offer a way out of the coordination problem. Here is the point where the concept of *risk dominance* (Harsanyi and Selten, 1988) kicks in. The risk is measured by the possible damage generated by unilateral deviations from the Nash outcomes. Take the angle of player 1. In particular, observe that playing t yields a payoff equal to 10 if the other player chooses l, but what if he/she chooses r? Parallel to this, also note that a unilateral deviation by player 2 from (b, r) gives player 1 a healthy payoff equal to 5. Player 1 of course dislikes the idea of getting zero, and is aware that a unilateral deviation by the rival is more harmful when departing from (t, l) than from (b, r).

Following Harsanyi and Selten (1972), we say that (b, r) risk-dominates (t, l) because

$$(10 - 5)(10 - 5) = 25 < 49 = (7 - 0)(7 - 0)$$

i.e., because the product of the losses generated by unilateral deviations is higher for (b, r) than for (t, l). The bottom line of risk dominance is to evaluate how attractive is unilateral deviation from any given Nash equilibrium of a game yielding more than one equilibrium in pure strategies. Or equivalently, risk dominance serves the purpose of comparatively assessing Nash equilibria in terms of their relative stability, and spot the one that emerges as the *largest basin of attraction*, so to speak, for the whole game.

3.5 Nash equilibrium in mixed strategies

Now we turn to probabilities. Theorem 3.1 establishes that every finite game yields at least one Nash equilibrium *in mixed strategies*. The mirror image of this claim is that there will be uncountably many finite games in which there exist no pure-strategy Nash equilibria. One such example is the so-called *matching pennies* game, appearing in Matrix 3.14.

Each player has two pure strategies, h ('heads') and t ('tails'). A quick look at the matrix reveals that this is a zero-sum game. This feature is not necessary to make the point, but it

Matrix 3.14 Matching pennies

$$
\begin{array}{c|cc}
 & & 2 \\
 & h & t \\
\hline
1 \quad h & 1;-1 & -1;1 \\
t & -1;1 & 1;-1 \\
\end{array}
$$

is usually adopted to clarify that the same applies as well to the subclass of constant-sum games that one would commonly solve via von Neumann's minimax (or maximin) criterion.

Along the main diagonal (where both agents play symmetrically), 2 pays one euro to 1, while the opposite happens along the secondary diagonal, in every asymmetric outcome. Hence, either 1 or 2 has a strict incentive to deviate unilaterally from every strategy combination, and the game has no Nash equilibrium in pure strategies.[14] However, according to Theorem 3.1, the equilibrium does exist in mixed strategies. Let's see how it works, and why.

To this end, suppose each player constructs a lottery by assigning complementary probabilities to each strategy, i.e., $\mathfrak{p}_{ih}$ and $\mathfrak{p}_{it}$, with $i = 1, 2$ and $\mathfrak{p}_{ih} + \mathfrak{p}_{it} = 1$, so that, for instance, player 1 may choose h with probability $\mathfrak{p}_{1h} = 2/5 = 0.4$ and t with probability $\mathfrak{p}_{1t} = 3/5 = 0.6$. Unlike what usually happens in lotteries, where probabilities are exogenously given, here it's up to players to set them. Therefore, the problem each player has to solve consists in choosing the probability vector *optimally*. What does *optimally* mean here? There are several different (but equivalent) ways of defining optimality in the context of a mixed-strategy equilibrium. One that – according to me at least – is rather intuitive, is the following: every player has to choose a vector of probabilities for him/herself, in order to make each of the rivals indifferent between her/his pure strategies. To understand why it must be so, consider what would happen if, say, for a given pair of probabilities $\mathfrak{p}_{ih} + \mathfrak{p}_{it} = 1$ attached by player i to h and t, respectively, player 2 were *not* indifferent between h and t. Being not indifferent, player j would have at least a weak preference for either one, i.e., either h or t would be weakly dominant over the other. The immediate implication would be that j would judge the game as being solvable in (at least weakly) dominant strategies, something that would create a logical paradox as, to begin with, the matching pennies game has no Nash equilibrium in pure strategies.

We may now apply this idea to Matrix 3.14, starting with the construction of the *expected payoff* for player 2, given a generic probability pair chosen by player 1. The expected payoff, indicated as $E\pi$, is nothing but the weighted average of the payoffs accessible to a player along any row or column, with weights being measured by the probabilities chosen by the opponent. Player 2 receives

$$E\pi_2(h) = -1 \cdot \mathfrak{p}_{1h} + 1 \cdot \mathfrak{p}_{1t}$$

by playing heads h, or

$$E\pi_2(t) = 1 \cdot \mathfrak{p}_{1h} - 1 \cdot \mathfrak{p}_{1t}$$

by playing tails t. Imposing the indifference condition $E\pi_2(h) = E\pi_2(t)$:

$$-\mathfrak{p}_{1h} + \mathfrak{p}_{1t} = \mathfrak{p}_{1h} - \mathfrak{p}_{1t}$$

The next step is the simple observation that $\mathfrak{p}_{1t} = 1 - \mathfrak{p}_{1h}$, as pure strategies are mutually exclusive. Therefore, the above equation can be rewritten as

$$1 - 2\mathfrak{p}_{1h} = 2\mathfrak{p}_{1h} - 1 \quad \text{or, equivalently,} \quad 2(2\mathfrak{p}_{1h} - 1) = 0$$

i.e., $\mathfrak{p}_{1h}^* = \mathfrak{p}_{1t} = 1/2$. By symmetry, also $\mathfrak{p}_{2h}^* = \mathfrak{p}_{2t} = 1/2$. The pair $(\mathfrak{p}_{1h}^*, \mathfrak{p}_{2h}^*)$ identifies the Nash equilibrium in mixed strategies – note that it is defined in the space of probabilities instead of in the space of pure strategies.

It is also interesting to revisit games with multiple equilibria in pure strategies, using the Nash equilibrium in mixed ones. For example, let's go back to the battle of the sexes, portrayed in Matrix 3.8. He (the row player) is indifferent between m and s if

$$\mathfrak{p}_{she,m} = 2 \cdot \mathfrak{p}_{she,s} \quad \Leftrightarrow \quad \mathfrak{p}_{she,m} = 2(1 - \mathfrak{p}_{she,m})$$

which yields $\mathfrak{p}^*_{she,m} = 2/3$ and $\mathfrak{p}^*_{she,s} = 1/3$. By the symmetry of the game along the main diagonal, we also have $\mathfrak{p}^*_{he,m} = 1/3$ and $\mathfrak{p}^*_{he,s} = 2/3$. This result deserves a few comments: first, both Nash equilibria will be realized with the same probability, equal to $\mathfrak{p}^*_{she,m} \cdot \mathfrak{p}^*_{he,m} = \mathfrak{p}^*_{she,s} \cdot \mathfrak{p}^*_{he,s} = 2/9$; second, 'mistakes' (either (m, s) or (s, m)) will happen with strictly positive probability, $\mathfrak{p}^*_{she,m} \cdot \mathfrak{p}^*_{he,s} = 4/9$ and $\mathfrak{p}^*_{she,s} \cdot \mathfrak{p}^*_{he,m} = 1/9$; third, the probability of making a mistake is higher than the probability of playing a pure-strategy Nash equilibrium, since $5/9 > 4/9$. This reinforces the general message generated by coordination games, about the need to introduce rules or laws to solve the problem and, ultimately, eliminate mistakes altogether.

A few words about the interpretation of the mixed-strategy Nash equilibrium are in order. This has been (and continues to be) a delicate issue for game theorists and all those who at some point have applied this solution concept to any given game in economics and the social sciences. After all, what does it mean that players will attach some well-defined probability vector to their pure strategies, if, when the time comes, they will indeed play a single pure strategy and neglect the others? Put differently, we may be happy to know that the matching pennies game admits a fully symmetric equilibrium in mixed strategies, so that any outcome is *a priori* equally likely to be played, but is this result any more informative than the fact that no equilibrium exists in pure strategies? An explanation often invoked relies on repetition and the laws of large numbers. If a game is repeated an extremely large number of times, then we might expect relative frequencies to reflect the vector of optimal probabilities identifying the Nash equilibrium in mixed strategies (or conversely, we may hope that probabilities will coincide with relative frequencies). However, to carry out exercises of this type, a game must be repeated for an enormous number of times, and this poses serious difficulties both empirically and experimentally.

An alternative route is to take a broader perspective and consider that we (social scientists) are not the first to encounter problems of this sort. The same discussions that are taking place among social scientists concerning the correct interpretation of a Nash equilibrium in mixed strategies as opposed to the pure-strategy equilibrium also occur among physicists as to the credibility of the quantum description of the world, as opposed to the Newtonian description. The latter discussion can be traced back to the famous dispute between Niels Bohr and Albert Einstein, who, while being one the fathers (although less than voluntarily) of quantum theory, claimed that 'God does not play dice with the universe'.[15]

As a last remark, I would like to draw your attention to a particular aspect of coordination games that has to do with probabilities. We have seen that one can invoke the concept of risk dominance as an equilibrium selection tool, by computing the product of deviation losses. Another way of performing the same exercise consists in making use of probabilities, as follows. Reconsider the game in Matrix 3.13, and suppose player 2 chooses l with probability $\mathfrak{p}_{2l}$ and r with probability $\mathfrak{p}_{2r} = 1 - \mathfrak{p}_{2l}$. With this in mind, player 1 conjectures to get either

$$E\pi_1(t) = 10 \cdot \mathfrak{p}_{2l} + 0 \cdot (1 - \mathfrak{p}_{2l}) = 10 \cdot \mathfrak{p}_{2l}$$

or

$$E\pi_1(b) = 5 \cdot \mathfrak{p}_{2l} + 7 \cdot (1 - \mathfrak{p}_{2l}) = 7 - 2 \cdot \mathfrak{p}_{2l}$$

The value of $\mathfrak{p}_{2l}$ – labelled as $\hat{\mathfrak{p}}_{2l}$ – such that $E\pi_1(t) = E\pi_1(b)$ is called the *risk factor* of the Nash equilibrium (t, l), while its complement to one, $1 - \hat{\mathfrak{p}}_{2l}$, is the risk factor associated with the Nash equilibrium (b, r). Elementary calculations show that $\hat{\mathfrak{p}}_{2l} = 7/12$ and therefore $1 - \hat{\mathfrak{p}}_{2l} = 5/12$. Fully symmetric considerations hold if one takes the standpoint of player 2. Interestingly, we can reconstruct the risk factors in terms of unilateral deviation losses, since

$$\hat{\mathfrak{p}}_{2l} = \frac{7 - 0}{(10 - 5) + (7 - 5)} = \frac{7}{12}$$

and

$$\hat{\mathfrak{p}}_{2r} = 1 - \hat{\mathfrak{p}}_{2l} = \frac{10 - 5}{(10 - 5) + (7 - 5)} = \frac{5}{12}$$

Either way, since $5/12 < 7/12$, (b, r) risk-dominates (t, l). More specifically, any $\mathfrak{p}_{2r} > \hat{\mathfrak{p}}_{2r}$ and $\mathfrak{p}_{1b} > \hat{\mathfrak{p}}_{1b} = \hat{\mathfrak{p}}_{2r}$ suffice to turn (b, r) into the risk-dominant equilibrium. This is a relevant message whenever players have no way of figuring out their respective probabilities, and therefore, say, agent 1 conjectures that $\mathfrak{p}_{2l} = 1 - \mathfrak{p}_{2l} = 1/2$. This automatically leads to select (b, r) as the risk-dominant equilibrium. To complete this rejoinder of risk dominance, observe that the pair $(\hat{\mathfrak{p}}_{1t}, \hat{\mathfrak{p}}_{2l})$ identifies the mixed-strategy Nash equilibrium of Matrix 3.13, that is, $(\hat{\mathfrak{p}}_{1t}, \hat{\mathfrak{p}}_{2l}) = (\mathfrak{p}_{1t}^*, \mathfrak{p}_{2l}^*)$, since $\hat{\mathfrak{p}}_{1t}$ and $\hat{\mathfrak{p}}_{2l}$ make player 2 indifferent between l and r and player 1 indifferent between t and b.

3.5.1 Subgame perfection: a further vindication

The subgame perfect equilibrium attained by backward induction is usually seen as a refinement of the Nash equilibrium concept. That is, subgame perfection is generally adopted as an instrument for equilibrium selection in games yielding multiple Nash equilibria in pure strategies. As a topping on the cake that I've offered you in this chapter, I will show you something more.

Examine the game illustrated in Matrix 3.15. Once again, strategies are t (for *top*) and b (for *bottom*) for 1, and l (for *left*) and r (for *right*) for player 2. For a moment, suppose it is played under imperfect information. If so, then this game possesses no Nash equilibria in pure strategies, and its solution requires the use of mixed ones (as a useful exercise, you can work them out).

If instead information is perfect, i.e., players move sequentially, the game in fact yields a subgame perfect equilibrium in pure strategies. This property can be verified by looking at Figure 3.7, where player 1 is at the root and therefore information is perfect for player 2.

Matrix 3.15

		2	
		l	r
1	t	1; 7	10; 0
	b	5; 1	4; 3

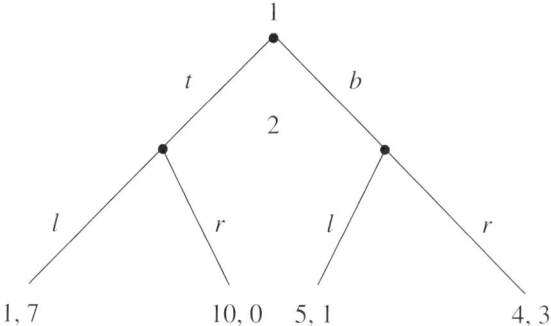

Figure 3.7 Perfect information for player 2

If 1 chooses *top*, 2's best reply consists in playing *left*. If instead 1 goes for *bottom*, then 2 prefers to go *right*. Hence, as 4 > 1, player 1 chooses *bottom* and the outcome (*b, r*) emerges as the subgame perfect equilibrium in pure strategies.

Further reading

The interested reader may find much more on the subject matter of this chapter in Fudenberg and Tirole (1991), Osborne and Rubinstein (1994) as well as in the other textbooks already mentioned at the end of Chapter 2. Classical and unsurpassed references for a thorough analysis of equilibrium refinements are Harsanyi and Selten (1988) and van Damme (1991). A comprehensive view on experimental games is offered by Roth (1993) and Camerer (2003), while a very nice and stimulating read is the paper by Goeree and Holt (2001).

3.6   Appendix: Schrödinger's paradox

Bluntly speaking, quantum physics can approximately be considered as a generalization of Newtonian 'classical' or deterministic physics.[16] In particular, when the phenomena being investigated involve (i) speeds that are considerably lower than the speed of light in vacuum, (ii) sufficiently low gravity and (iii) objects that are neither too small nor too big, then one can proceed according to the standard Newtonian model. Yet, when it comes either to the very basic particles (e.g., subatomic particles like electrons and sub-subatomic components like quarks), or to the behaviour of objects characterized by an extremely high mass and therefore also large gravitational force (e.g., black holes, quasars and pulsars), quantum features become so relevant that they cannot be disregarded.

To understand these seemingly puzzling aspects of physics, one has to accept the idea that our capability to investigate the world of subatomic particles is inherently affected by *indeterminacy*. This is the so-called *Heisenberg's law*, whose *vulgata* can be spelled out intuitively in the following terms: We cannot know at the same time how fast a subatomic particle (say, an electron) is moving, *and* where it is exactly. The knowledge we can produce is confined to a distribution of the probability that, at any given time, such a particle will be within a certain interval, given its speed.

As an illustrative example, consider the story traditionally known as *Schrödinger's paradox* or *Schrödinger's cat in the box*, which I am about to tell according to a *vulgata* commonly adopted in the current literature in the field.[17]

This paradox refers to an experiment, where a cat is locked inside a box, in front of a handgun, which is loaded.[18] The box is opaque to both light and sound, so that the physicist must open it to observe the cat's condition. The trigger is linked to a sensor, located outside the box, in front of a bulb. Assume the physicist can switch on the bulb, producing a single quantum of light (a photon) at a time. The sensor measures the spin of the photon. In the remainder, I will assume conventionally the following. If the photon rotates rightwards (i.e., the spin is positive), then the sensor pulls the rope (and the trigger), and the handgun shoots the cat dead. If instead the photon rotates leftwards (i.e., the spin is negative), then the sensor does not pull the rope (and the trigger), so that the cat survives.

The paradox takes place in the physicist's mind before the experiment is carried out, or, equivalently, before he/she opens the box to see whether the cat is still alive or not. The physicist knows that, *ex post*, the cat is going to be either dead or alive. However, *ex ante*, the two states of the system–cat (i.e., *alive* and *dead*) as well as the two states of the world (which can be labelled as *handgun fired because spin was positive* and *handgun didn't fire because spin was negative*) coexist in the experimenter's mind. This paradox persists until he/she opens the box to observe the cat's state of health.

Of course, we cannot observe the spin of a single photon. However, if we take a very large number of photons, and measure their spin with appropriate instruments, we will come up with the information that half of them rotate rightwards (leftwards). Hence, when a single quantum of light – randomly drawn from that large number – comes into play, our prior (as well as the experimenter's) is that it will rotate either way with probabilities $p_l = p_r = 1/2$. Accordingly, the handgun will shoot the cat dead with probability $1/2$ (otherwise, again with probability $1/2$, the cat will stay alive). Now define (i) the event 'shooting' as s and the alternative 'not shooting' as ns; and the possible states of the cat as a for 'alive' and d for 'dead'. Given these probabilities, the experimenter may forecast the state of the cat at the end of the experiment according to the following expression, that physicists call the *density matrix*:

$$D = \frac{1}{2}\langle d \mid s \rangle + \frac{1}{2}\langle a \mid ns \rangle \qquad (3.3)$$

which says that our expectations about the state of the cat is that it will be dead or alive with equal probabilities, given that the gun has or hasn't fired. The symbol | stands for 'given that' or 'conditional upon'. The peculiar 'angled parentheses', $\langle$ and $\rangle$, appearing in (3.3) are known as *bra* and *ket*, respectively, after Dirac (1930), who was the first to adopt them. The introduction of the density operator can also be traced back to Dirac (1930) and pops up again in von Neumann (1932). This indicates that the isomorphism between (i) the theory of expected utility, game theory and contemporary mathematical economics in general, on the one hand, and (ii) quantum mechanics, on the other, can hardly be considered as accidental.

My personal opinion is that, in *Theory of Games and Economic Behavior*, von Neumann and Morgenstern (1944) made use of a toolkit borrowed from Dirac's formalization of quantum mechanics,[19] without bothering to make this fact explicitly known either to economists or to physicists.

The density matrix (3.3) can be rewritten in a form that will look somewhat more familiar to us, to clarify the analogy with what economists would call *expected value*. Indeed, a social scientist would write the expected state of the cat (c) as follows:

$$E(c) = \frac{1}{2}(d \mid s) + \frac{1}{2}(a \mid ns) \qquad (3.4)$$

Notice that the above reads the same in both disciplines, independently of whether we write it as in (3.3) or as in (3.4), i.e., 'with probability 1/2 I will observe a dead cat because the cat is in fact dead (or equivalently, because the handgun shot, as the spin was positive), and with the same probability I will observe that the cat is still alive (because the handgun did not shoot, as the spin was negative)'. Observe that the meaning of the symbol | is exactly the same in both cases.

As soon as the experimenter opens the box to observe the state of the system, probabilities collapse to either zero or one, depending on whether the gun shot the cat or not – and this will reveal *ex post* the spin of the photon. Consequently, the experimenter observes either *ca* or *cd*, but surely not both states at the same time. On the contrary, as long as the experimenter does not open the box, the two states *coexist in a quantum sense*.[20]

To conclude, I would like to stress that, when economists build up the expected value of an agent's payoff, they face exactly the same kind of paradox that I have illustrated above concerning *Schrödinger's cat in the box* parable. Hence, also in economics the issue is the transition from the *ex ante* (quantum) prevision to the *ex post* (Newtonian) observation. In both settings, the limit to our knowledge is inherently given by the indeterminacy associated with (i) the measurement in physics and (ii) the (random) transition from a mixed strategy to a particular pure strategy in game theory.

4 Understanding economics

I've been spending my money
In the old town
It's not the same honey
When you're not around

Philip Lynott,
Old town

Well, I haven't written this chapter to offer you an interpretation of compulsive shopping – which is anything but strategic, right? – but I may take Phil Lynott's lyrics as a sort of oxymoronic starting point for talking about economic incentives governing market behaviour. And, since economics is the area of the social sciences that was the first to massively adopt game theory, I will use it as a springboard from which I will dive (and you with me) into the wide sea of applications we are going to run across throughout the remainder of the book. An exhaustive survey of all the possible applications of game theory to economics is well beyond the scope of the present volume. However, I will strive to offer you a flavour of what has been and is being done in this field, so that, hopefully, you may desire to know more, and extend your reading outside the narrow boundaries of this book.

The structure of this chapter is threefold. Section 4.1 covers industrial economics, and illustrates the game-theoretic approach to price competition, innovation incentives, product differentiation and advertising, and entry barriers. These are classical topics that have been intensively investigated with the tools of game theory. As a topping on the cake of industrial organization, the comparatively newer issues of standardization, compatibility and network externalities are also briefly dealt with. Section 4.2 is devoted to the analysis of some strategic aspects of macroeconomics and macroeconomic policy, such as rules versus discretion, and the coordination of monetary and fiscal policies. Last, but not least, Section 4.3 investigates the basics of the economics of natural resources and the environment.

4.1 Industrial economics

4.1.1 Prices and market shares in oligopoly

In accordance with industrial economics, we may classify a market in terms of the degree of power enjoyed by the firms operating in it. By definition, holding monopoly power is the best of all possible perspectives for any firm, as it implies that it will operate alone in the market, without competitors.[1] The opposite case is perfect competition, the Nirvana of economic analysis, where infinitely many atomistic firms supply the market without the slightest degree of control on price. In both cases, proper strategic interaction is absent. In between, there's oligopoly: a market with a relatively small number of firms that are aware of each other and exert some control on prices and market shares.

Oligopoly theory dates back to the French mathematician Cournot (1838). His model, in its original version, is still the foundation of any course in industrial economics in schools across the world. Cournot tells a story about two millers selling the same type of flour and choosing simultaneously their respective production levels so as to maximize individual profits, with the price being determined by the demand side of the market. In the modern jargon of game theory introduced in Chapter 2, it is a non-cooperative variable-sum game with imperfect, complete and symmetric information.

Almost half a century later, another French mathematician, Bertrand (1883) criticizes a specific aspect of the work of Cournot, namely, the idea whereby firms decide quantities or market shares. If this is the case, who sets the price? Shall we seriously think that the demand side (i.e., consumers) will decide at what price to buy the firms' products? Observation suggests that this is (almost) never the case, says Bertrand. Thence the opposite assumption is made that firms set prices and let consumers decide how much to buy on the basis of the extant price lists. Much later, this critique has given rise to a sort of legend according to which there would exist a *Bertrand model*, as opposed to the Cournot model. This is indeed no more than an *ex post* artifact built up *ad hoc* on the basis of the opinion expressed by Bertrand himself in his 1883 article.[2]

Be that as it may, let's see how price competition works in a basic duopoly model. Consider a market supplied by two firms that use the same technology and sell exactly the same homogeneous good (i.e., identical soft drinks, sweaters, or cars). These assumptions are admittedly unrealistic – and I will relax both of them in due course – but are initially adopted for the sake of simplicity and also because they allow one to single out some basic properties of oligopolistic interaction.

Let's say that firms, labelled as 1 and 2, produce at the same unit cost c. The range of admissible prices is $[c, p_M]$, p_M being monopoly price. Now suppose firm 1 sets a price $\hat{p}_1 = c + \eta$, with $\eta > 0$ measuring the mark-up (the unit profit), and such that $\hat{p}_1$ belongs to the admissible range. If the solution concept is the Nash equilibrium, is it optimal for firm 2 to set the same price, or is there a different best reply to $\hat{p}_1$? Quite intuitively, firm 2 will find it optimal to choose a price $\hat{p}_2 = \hat{p}_1 - \varepsilon = c + \eta - \varepsilon$, with ε positive but arbitrarily small (say, one cent) in such a way that:

- $\hat{p}_2 \geq c$, and therefore firm 2's profits will be positive; and
- all consumers willing to buy at $\hat{p}_2$ will buy from firm 2 only, since the product is homogeneous and therefore what matters is just the (slightest) price difference.

However, $(\hat{p}_1, \hat{p}_2)$ cannot be a Nash equilibrium: indeed, given $\hat{p}_2$, it will be optimal for firm 1 to deviate from $\hat{p}_1$ towards $\tilde{p}_1 = \hat{p}_2 - \varepsilon$, as long as ε is sufficiently small to ensure that $\tilde{p}_1 \geq c$. At this point firm 1 is enjoying monopoly power over this market, but again this cannot be an equilibrium because a unilateral downward deviation is optimal for firm 2, and so on and so forth … until both firms price at marginal cost, with $p_1 = p_2 = p^{BN} = c$. It is easily checked that no unilateral deviation from marginal cost pricing is profitable: deviating below marginal cost would yield monopoly power, but profits would be negative; deviating above marginal cost is pointless as the deviator would have no customers at all.[3] Hence, this is the unique Bertrand–Nash equilibrium of the duopoly game with homogeneous goods and symmetric technologies. The bottom line of the story I've been telling you is summed up in the following remark.

REMARK 4.1 With homogeneous goods and identical technologies, the undercutting mechanism leading to marginal cost pricing jeopardizes firms' profits altogether and replicate the same equilibrium outcome that one would expect to appear under perfect competition.

This striking (and quite peculiar) result deserves a closer look. The source of marginal cost pricing at the Bertrand–Nash equilibrium is a price war driven by the lust for monopoly power. This is the prize that both firms are looking for when they undercut each other's price – sadly enough, by doing so they destroy the prize itself, and generate its opposite. This means, of course, very good news for consumers, who may initially worry about the undesirable consequences of strategic interaction among a small number of firms, and ultimately enjoy consumption at the lowest possible price, as in their wildest dreams. For these reasons, the Nash equilibrium outcome of the Bertrand game is traditionally known as the *Bertrand paradox*. From the 1950s, this model has been considered as a theoretical explanation of the self-regulating ability on the part of any given market, and it has been adopted as a benchmark by the Chicago School in claiming that 'two is enough for competition'. Note that the extension to any number of firms does not modify the picture, as long as all of them use the same technology (with the same unit cost) to supply the very same object.

This generates a fundamental question: is the Bertrand paradox *credible*? Or, is it *realistic*? A casual look at the game reveals immediately several weaknesses. To begin with, in the real world it is almost never the case that firms supply perfect substitutes. Goods are almost never homogeneous, and some degree of differentiation meeting the heterogeneous tastes of consumers is systematically adopted by firms precisely to soften price competition. And even if tastes were the same across consumers, it would be a smart idea for firms to use intense advertising campaigns to modify consumers' preferences so as to persuade some consumers that they like a product they have never thought about before.

Additionally, I would like to draw your attention to another possible modification of the initial Bertrand set-up. It is small and obvious, but its consequences are extremely relevant and open completely new perspectives on the way we may look at firms' behaviour and incentives. Simply suppose the two firms have different unit costs, with $c_2 > c_1$. That is, firm 1 is more efficient than firm 2. If so, then any price $p_1 \in (c_1, c_2)$ grants firm 1 monopoly power with strictly positive profits. In fact, the optimal monopoly price in this situation will be $p_1 = c_2 - \varepsilon$, because then firm 2 cannot undercut firm 1 without incurring a loss. Again, this result is also replicable for any number of firms larger than two: the most efficient one in the industry will surely become a monopolist. Price competition melts away in the face of any technological gap across any population of firms selling the same good and using prices as their market variables. This prompts for the analysis of the sources of technical progress, which is exogenous to the Bertrand model as such, but may cause the latter to collapse into a monopoly, i.e., into something that is no longer a game. The relationship between market power and innovation incentives is a long-standing issue in industrial economics, with many relevant implications in terms of policy and economic growth. I'll come back to these extensions in the remainder of this chapter. Again on costs, a serious drawback of the Bertrand game is that firms are supposed to operate without capacity constraints; or, if these are present, they are so innocuous that the cost of building up productive capacity is not even accounted for, and what matters is the unit cost only. If, a bit more realistically, each firm has a plant of size k and has to pay a rental price of capital r per unit of installed capacity, then pricing at $p^{BN} = c$ would indeed kill both firms, as neither of them would be able to pay the cost of capacity rk.

Another delicate issue is the following. The Bertrand model is a one-shot game replicating perfect competition because of a price war that firms would possibly learn to avoid by taking into consideration that their interaction will repeat over time, possibly over an arbitrarily long time horizon. The ultimate outcome of such awareness is a harmful cartel with firms replicating the behaviour of a monopolist – exactly the opposite of the reassuring Bertrand paradox. This is the topic treated in the next chapter, where I will also briefly illustrate the essence of the Cournot (1838) model.

To conclude this overview of the price competition oligopoly game *à la* Bertrand, it is interesting to note that it can be interpreted as an *auction*. Usually, the standard mechanism observed at auction houses like Sotheby's or on eBay is that of ascending bids. The object being auctioned goes to the agent who has placed the highest bid. Depending upon the rules of the auction, the winner may or may not pay that highest bid. In the so-called first-price (or English) auction, the winner indeed pays the highest bid he/she has placed. In the second-price (or Vickrey[4]) auction, he/she will only pay the second highest bid, plus possibly a step included at the outset in the auction rules and known *a priori* to everybody (like on eBay). The Bertrand game looks pretty much like an auction with descending bids: firms offer descending prices to win the object of the auction, namely, monopoly power over the market. Depending on the cost structure of the industry, this auction will be a first- or second-price one. Typically, with fully symmetric firms operating at the same efficiency level, the game is a first-price auction, where no firm is able to win because of the paradox, while any asymmetric game with different efficiency levels is a second-price auction, where the firm endowed with the most efficient technology gets monopoly power. Auctions will prove very useful in explaining several phenomena we will run across in this chapter as well as in Chapter 7.

4.1.2   *Innovation and advertising*

Economic growth is largely driven by technical progress generated by investments in R&D, and therefore the analysis of firms' incentives to generate innovations is a very relevant task with the highest priority on the industrial economists' agenda.

The output of R&D activities can be either a *process* innovation or a *product* innovation. Process innovation identifies a form of technical progress whereby firms continue to produce the existing goods, but do so at a lower cost. Instead, product innovation yields new varieties of existing goods, or completely new commodities that consumers have never seen before, given the existing technology. Of course, common sense suggests that, in many (or most) cases, the two types of innovation pop up hand in hand. One such example is the introduction of compact disk (CD) players, which replaced turntables worldwide in the early 1980s: a CD player is a new product that incorporates a new technology with its own specific production costs. Yet, for the sake of analytical tractability, most of the extant analysis of this issues in modern industrial economics treats the two types of innovation separately. In this section, I will give you a flavour of both, one at a time, and also, tentatively, a sketch of their interplay in a single model.

Comparatively speaking, the first and foremost aspect that has attracted the attention of professional researchers in this area is the relationship between market power and innovation incentives. The related debate has been triggered by Schumpeter (1934, 1942) with the so-called *Schumpeterian hypothesis*, whereby innovation incentives are increasing in the degree of market power: R&D activity needs funding, so what could possibly outperform monopoly in this respect? The cornerstone of the Schumpeterian argument is that a monopolist can do at least as well as any other market form, that is, the profits of a firm standing alone on the marketplace are at least as high as the *collective* profits of any number of oligopolistic or perfectly competitive firms. Quite compelling, isn't it? So much so that the Schumpeterian claim remained unchallenged for a long time.

Later, Arrow (1962) proposed an opposite view of the matter, based on the idea that small firms operating in competitive environments could actually strive harder than a monopolist to attain any given innovation. The bottom line of the Arrovian position can be summarized as follows. Let's suppose there exists a patent law granting an innovator the property rights on a new product or technology, as happens in virtually any economic system on the planet. A perfectly competitive firm is small, gets zero profits by definition and is just one out of a large number of similar productive units. Hence, it has no internal funding sources. However, it can get indebted with banks to raise the necessary resources for R&D, on the basis of the prize accruing to the winner of the innovation race possibly involving a large number of competitors. The prize is the flow of profits the winner will obtain after patenting the innovation. Now note that a perfectly competitive firm patenting an innovation jumps abruptly from zero to monopoly profits, while a monopoly firm patenting the same innovation just replaces itself. This feature of the problem, which is known as the *replacement effect*, speaks in favour of small competitive firms and against monopolistic ones, and therefore the Arrovian position is that the former should be more innovative than the latter. A simple game, based upon a modern revisitation of their debate (Gilbert and Newbery, 1982 and Reinganum, 1983), can be constructed to illustrate the two opposite views of Schumpeter and Arrow.

To simplify the picture, there will be only two players: a monopolist and a potential entrant, the latter firm being allowed to enter the market if and only if it gets the innovation. Also assume, for instance, that the innovation at stake takes the form of a new product that does not replace completely the existing one. An appropriate example could be that of solid-state amplifiers for domestic hi-fi sets, which were introduced in the late 1960s without eliminating tube amps.[5] Before the innovation is adopted by anyone, the market is a monopoly with the incumbent selling the 'old' product, the resulting profits being $\pi_M(\text{old})$. The innovation game can be illustrated as an auction. A third subject (the real innovator) is a genius (say, Albert Einstein) with many striking ideas but no productive facilities at all. Hence, having patented the new product, Albert decides to auction the patent, the monopolist and the potential entrant being invited to bid.

The maximum bid the incumbent can place is measured by the amount of monopoly profits it can attain after winning the auction by marketing the new product together with the old one, $\pi_M(\text{old} + \text{new})$. If the incumbent does not win the auction, it gets a duopoly profit $\pi_D(\text{old})$ by continuing to sell the old product. On the other hand, the maximum bid the outsider can afford is the amount of duopoly profits generated by the innovation, $\pi_D(\text{new})$. According to Gilbert and Newbery (1982), the comparative incentives of the two firms are summarized by the following inequality:

$$\pi_M(\text{old} + \text{new}) - \pi_D(\text{old}) > \pi_D(\text{new}) \tag{4.1}$$

or, equivalently,

$$\pi_M(\text{old} + \text{new}) > \pi_D(\text{old}) + \pi_D(\text{new}) \tag{4.2}$$

Condition (4.1) states that the incumbent will outbid the potential entrant (and therefore monopoly will persist) because the monopoly profits generated by the innovation – joint with the old good – are larger than the sum of duopoly profits with one firm supplying the old product and the other supplying the new one – which appears explicitly in (4.2). In a nutshell, this is the Schumpeterian interpretation of an innovation game: given that a monopolist can always at least replicate the performance of any industry with at least two firms, at the Nash equilibrium of the auction the incumbent outbids the entrant and the market remains a monopoly.

The Arrovian version relies on the fact that there is an alternative and equally plausible way of measuring firms' relative incentives to bid (or to invest). According to Reinganum (1983), one should work out the game anew using the same basic elements (profits), in such a way that either firm's incentive is given by the difference between *ex ante* and *ex post* profits. The profit increase accruing to the incumbent replacing itself is $\pi_M(\text{new}) - \pi_M(\text{old})$, while that accruing to the potential entrant starting from scratch is $\pi_D(\text{new}) - 0$. Comparing these incremental profits yields:

$$\pi_M(\text{old} + \text{new}) - \pi_M(\text{old}) \overset{?}{\gtreqless} \pi_D(\text{new}) \tag{4.3}$$

and inequality (4.3) can take either sign, depending on the 'size' of the innovation (for instance, product quality, durability, etc.).

It is worth stressing that the left-hand side of (4.3) is a direct measure of the replacement effect, and *may* speak in favour of the entrant. If it does, then the incumbent rests on its laurels; as a result, the Nash equilibrium of this auction game is that the entrant outbids the incumbent and the former monopoly becomes a duopoly.

The same framework can be quickly adapted to illustrate the incentives for process innovation, in the form of a decrease in unit production costs. Players are the same as above, the incumbent, a potential entrant, and our friend Albert. The product offered to consumers is homogeneous, but its unit cost can be reduced from $\bar{c}$ to $\underline{c} < \bar{c}$ via a non-drastic innovation auctioned by Albert. The Schumpeterian view *à la* Gilbert and Newbery (1982) is summarized by two conditions, which replace (4.1) and (4.2):

$$\pi_M(\underline{c}) - \pi_D(\bar{c}) > \pi_D(\underline{c}) \tag{4.4}$$

and

$$\pi_M(\underline{c}) > \pi_D(\bar{c}) + \pi_D(\underline{c}) \tag{4.5}$$

Once again, monopoly persists because the monopoly profits generated by the new technology are strictly higher than the total profits of a duopoly in which one firm is endowed with an inferior technology.

The Arrovian perspective *à la* Reinganum (1983) is instead based on the equivalent of (4.3):

$$\pi_M(\underline{c}) - \pi_M(\bar{c}) \overset{?}{\gtreqless} \pi_D(\underline{c}) \tag{4.6}$$

whose sign may indeed flip over depending on the difference between old and new unit production costs.

We can sum up all this in the following remark.

REMARK 4.2   The relationship between market power and innovation is ambiguous. If the incentive is measured by the profit perspective associated to the innovation, big firms will systematically outperform small ones in R&D races. If instead the incentive is based on incremental profits, then the opposite *may* apply.

A relevant corollary to this discussion is the role of patent laws. It is commonly recognized that a patent law is needed to safeguard the firms' incentives to carry out R&D investments, as in the absence of patentability any innovation could be immediately copied without legal consequences – the only consequence being that the innovating firm would go bankrupt as it would be unable to cover the cost of the investment. Hence, the monopoly rights granted by any patent law are seen as a reward for innovation and a stimulus to technical progress.[6] Yet, opinions are not unanimous on this point, for more than one reason. First of all, antitrust laws speak clearly against monopoly power. When we see how cartels operate (in Chapter 5), this issue will pop up again in more detailed terms. Second, the design of patent coverage is a delicate exercise, centred upon the most delicate trade-off between static and dynamic efficiency. In summary, static efficiency is attained when the welfare generated by an industry is maximized, and this happens when – bluntly speaking – all firms may access the same knowledge and technology and the degree of competition is satisfactorily high. Ideally, it is so high that the equilibrium price falls to marginal cost and the industry is perfectly competitive. If we want an economy to attain dynamic efficiency through a fast innovation pace, we have to accept the idea that monopolies will prevail for non-negligible time spans after any innovations have been patented. Shortening the patent length hinders (respectively, fosters) dynamic (respectively, static) efficiency, and conversely. That is, we cannot get two eggs in one basket. Third, the protection of property rights might be in open contrast with our concern about the health and ultimately the survival of entire nations. The possibility of allowing innovations to be copied freely is a hot issue in the pharmaceutical industry, in particular when it comes to antiviral drugs – for instance, for the treatment of HIV. Is the standard property right argument a sound one, in this respect? I let you judge the matter for yourself.

R&D activity poses some serious policy issues that governments have to deal with carefully. One such aspect is the possibility of subsidizing firms' innovative efforts using public funds.[7] This matter can be illustrated by resorting to a 2×2 game between two firms, 1 and 2, as in Matrix 4.1, describing the situation arising in the absence of subsidies to either firm. Strategies c and nc stand for *to compete* and *not to compete*, respectively.

The game has a 'chicken-like' structure, with two asymmetric Nash equilibria along the secondary diagonal (and of course also an equilibrium in mixed strategies). Therefore, *a priori*, one remains in doubt as to the actual outcome of strategic interaction. What if one of the two firms has an advantage over the rival, perhaps due to a subsidy from its government? Suppose this happens to firm 2, which receives an amount equal to 15 from its government in

Matrix 4.1 The game without subsidies

		2	
		c	nc
1	c	$-10; -10$	$30; 0$
	nc	$0; 30$	$0; 0$

Matrix 4.2 The game with subsidies to firm 2

		2	
		c	*nc*
1	*c*	−10; 5	30; 0
	nc	0; 45	0; 0

Matrix 4.3 The game with symmetric subsidies

		2	
		c	*nc*
1	*c*	5; 5	45; 0
	nc	0; 45	0; 0

order to compete against firm 1 (if firm 2 decides not to compete, then the subsidy vanishes: therefore firm 2's payoff is modified only along the first column). Consequently, the game takes the form of Matrix 4.2.

The presence of a subsidy entails that *c* becomes a dominant strategy for firm 2, and naturally firm 1 is aware of that. The resulting process of iterated deletion of dominated strategy selects (nc, c) as the unique equilibrium of Matrix 4.2. That is, the strategic advantage artificially created through the subsidy policy drives the game towards an equilibrium where the subsidized firm acquires monopoly power. Of course firm 1 may invoke the adoption of an analogous measure on the part of its own government, whereby both firms' activities are equally subsidized with public money. The outcome is illustrated in Matrix 4.3.

This game can be solved in dominant strategies, yielding (c, c) as the unique equilibrium – which, by the way, is also Pareto-efficient from the firms' viewpoint. Whether this holds as well for both countries at an aggregate is a delicate issue, as the answer requires a careful evaluation of the pros and cons of the subsidy policy in terms of social welfare. For more on this side of the issue, see Krugman and Obstfeld (1988).

4.1.3   Product differentiation

> You can have it any color you want,
> as long as it's black
>
>   Henry Ford

We have just seen that, in the extreme case of homogeneous goods, price competition drives profits to zero. Hence, there arises an obvious incentive for firms to differentiate their goods in order to sustain prices above unit costs. This is a comparatively old idea in industrial economics, dating back at least to Hotelling (1929). Here I will lay out a simple game inspired to the original Hotelling model and its updated version due to d'Aspremont *et al.* (1979).

The space of all possible product varieties is a linear segment whose length can be set equal to one without loss of generality. The segment, for instance, may measure the amount of sugar in a generic soft drink. Along the segment is uniformly distributed a population of consumers, one at every point along it. The position of each consumer reveals his/her preferences as to the main feature of the good being traded in this market (sticking to the aforementioned example, the consumer located at 0 prefers Cola Zero, while the consumer at 1 prefers the usual one). The total mass of consumers is therefore equal to one. Two

Figure 4.1 The Hotelling segment

single-product firms, 1 and 2, compete in this market, in the following way. Firms have the same technology, again measured by the symmetric unit cost c. They have to choose their respective locations x_1 and x_2 non-cooperatively, under complete, symmetric and imperfect information (i.e., simultaneously) knowing that market coverage is full (that is, every consumer will buy a can of soft drink) and the set of admissible locations is $\{0, 1/2\}$ for 1 and $\{1/2, 1\}$ for 2. Moreover, firms also know that the price (as well as the resulting price–cost margin per unit) is increasing in the amount of product differentiation $x_2 - x_1$:

$$p_1 = p_2 = c + \theta(x_2 - x_1) \tag{4.7}$$

The above pricing rule, joint with firms' locations, will determine market shares and profits. Let's look at each admissible outcome in turn. The first possibility is that both firms locate in the middle of the segment, with $x_1 = x_2 = 1/2$. This causes both prices to fall upon unit cost c because product differentiation is absent altogether, and therefore profits will be nil for the very same reason as in the Bertrand game. The second scenario is one in which firms locate at the opposite ends of the spectrum of varieties, with $x_1 = 0$ and $x_2 = 1$. In such a case, the resulting prices will be $p_i = c + \theta$, $i = 1, 2$. Market demand will split evenly as a consequence of maximum differentiation: all consumers preferring low-sugar drinks – those located between 0 and $1/2$ – will patronize firm 1, while those located between $1/2$ and 1 will buy from firm 2 (the consumer located at $1/2$ will toss a coin to decide). The resulting profits are again symmetric but positive, amounting to $\pi_i(0, 1) = \theta/2$. The last case is that in which one firm locates at $1/2$ and the other locates at one of the endpoints, either 0 or 1. Suppose firm 1 chooses $x_1 = 0$ and firm 2 chooses $x_2 = 1/2$. Prices will be $p_i = c + \theta/2$. Now note that every consumer between $1/2$ and 1 will buy from firm 2, as will all of the consumers between $1/4$ and $1/2$, because all of them prefer variety 2 to variety 1 at equal prices. Only those located between 0 and $1/4$ will become firm 1's customers – the consumer at $1/4$ will be indifferent. This asymmetry in market shares determines the following profits:

$$\pi_1\left(0, \frac{1}{2}\right) = \frac{\theta}{8}, \quad \pi_2\left(0, \frac{1}{2}\right) = \frac{3\theta}{8} \tag{4.8}$$

The opposite case where firm 1 locates in the middle and firm 2 locates at 1 yields:

$$\pi_1\left(\frac{1}{2}, 1\right) = \frac{3\theta}{8}, \quad \pi_2\left(\frac{1}{2}, 1\right) = \frac{\theta}{8} \tag{4.9}$$

with the indifferent consumer at $3/4$. The linear space of product characteristics with the three admissible firms' locations is represented in Figure 4.1.

The outcome of the strategic interplay between firms can be investigated observing the 2×2 in Matrix 4.4. The game is symmetric along the main diagonal, so that we may take

Matrix 4.4 Product differentiation
in the Hotelling game

		2	
		1	1/2
1	0	$\theta/2; \theta/2$	$\theta/8; 3\theta/8$
	1/2	$3\theta/8; \theta/8$	$0; 0$

either firm's viewpoint knowing in advance that an analogous interpretation will apply to the rival as well.

Given that $\theta/2 > 3\theta/8$ and $\theta/8 > 0$, the outcome $(0, 1)$ emerges as the unique Nash equilibrium of the game, at the intersection of strictly dominant strategies. The associated degree of differentiation is maximum, with prices well above unit production costs. This result, due to d'Aspremont *et al.* (1979), is one of the cornerstones of the modern theory of industrial organization, and can be understood by looking at the interplay between two opposite forces driving firms' location choices. The first is labelled as the *competitive* or *strategic effect*: firms strive to increase product differentiation, as this softens price competition. The second is the *demand effect*: each firm would like to capture the preference of the median consumer at $1/2$ (the 'middleman'), as this expands considerably the firms' market share, all else being equal. Overall, the balance speaks in favour of the strategic effect, which plays a major role and shapes the equilibrium configuration of the industry so as to yield maximum differentiation.

The above discussion can be usefully summarized in the following remark.

REMARK 4.3 Product differentiation softens price competition. Hence, firms endowed with analogous technologies will exploit product differentiation in order to sustain prices above unit production costs.

Note that the strategic choice of product varieties is not taken to meet consumer preferences. It's actually the other way around: any heterogeneity in consumer preferences will be exploited by firms to extract surplus from consumers. In this respect, an even sharper comment could be that, if consumer preferences are not heterogeneous, then firms may use advertising to artificially create some profitable heterogeneity.

This model can be modified in several directions to account for specific features of industries and strategic interactions among firms operating in such industries. A useful extension is loosely based on Gallini (1992). What if firms enter the market sequentially, perhaps due to sequential innovations? First of all, relax the above assumption concerning admissible locations, and admit instead that $x_i \in \{0, 1/2, 1\}$ for both firms. Suppose firm 1 innovates first. Say, firm 1 is DrinkThis, which can supply the market with a single variety of its soft drink, under full market coverage. If it anticipates the entry of a competitor – TasteThat – at a later date, DrinkThis will choose to locate at one of the endpoints in order to allow the second firm to differentiate its product from 'Fizz' and avoid tough price competition. In fact, the earliest variety of Fizz is full of sugar; therefore, we may set $x_{DT} = 1$, the subscript DT standing for DrinkThis. As everybody knows, Coca Cola never patented Coke, its exact recipe being 'protected' through secrecy. This is a crucial feature that deserves a careful explanation. Given that any patent elapses at some point, existing products can indeed be copied by competitors as soon as they are no longer protected by patents, and the accessibility of deposited patents makes any product's feature publicly known as soon as the patents are conceded to innovators. If, instead, a product is never patented, and the veil of secrecy around it is thick enough, competitors may never get to know exactly how it is being manufactured. The downside of secrecy is that, if rival firms can work out a more or less perfect copy, they can market it without any legal consequences.

Having said that, we may turn our attention to the strategy of TasteThat, which may face two possible scenarios: the first is one where secrecy doesn't work, and TasteThat successfully discovers the correct proportions of ingredients in Fizz; the second is the opposite, where the veil of secrecy has remained an impassable barrier for TasteThat, whose only alternative is to market a differentiated soft drink. If TasteThat cannot discover the exact recipe used by DrinkThis, then obviously the only equilibrium entails maximum differentiation. If instead TasteThat unveils the secrets of their rival, then the crucial question is this: Provided one is able to copy the rival's goods, is it wise to copy a successful product? Judging from the Hotelling model, the answer is 'no', as imitation triggers a price war that ultimately drives profits to zero. This, however, is conditional on the relative productive efficiency of the two firms. If the unit production costs of DrinkThis and TasteThat are exactly the same, then the sequential entry game along the Hotelling segment has a unique equilibrium in pure strategies, with DrinkThis entering first at 1 and TasteThat subsequently locating at the opposite endpoint. If instead the time interval is fruitfully exploited by TasteThat to carry out some R&D effort for process innovation lowering the unit production cost, then TasteThat will indeed profit from locating at 1, replicating Fizz exactly and stealing all of DrinkThis's customers with an identical product sold at a slightly lower price.

Advertising campaigns are very familiar to all of us, at least superficially. Firms happen to invest huge sums in ads on the TV, in magazines, newspapers and on the web. Why do they? OK, the obvious answer is that they want to convince consumers to purchase their products. But this prompts other and more subtle questions that need accurate and less obvious answers. Should we take the messages conveyed by ads at face value? That is, are firms systematically telling the truth about their products (and their rivals' products)? Is the amount of money invested in advertising by any given firm or industry too high or too low as compared to what would be socially optimal? Advertising can be either informative or persuasive. It is informative when it tells consumers that a given product exists, has a given set of features and is sold at a certain price. It is persuasive when it is aimed explicitly at convincing a consumer already informed about the product features to buy such a product, or to buy it at a higher price. It goes without saying that, quite often, ads are both informative and persuasive at the same time. Additionally, advertising may also be comparative, something like 'buy my product because it's better than the rival firms' alternatives' for some reasons. This type of advertising is often prohibited – although it is not in electoral campaigns (see Chapter 6).

An advertising game based on the Hotelling model can be usefully illustrated here. Firms, by assumption, are in 0 and 1. The game is one in which advertising modifies market shares, all else being equal (including prices, corresponding to $p_i = c + \theta$ as above). Let's say that advertising entails a fixed cost F, so that: (i) if both firms invest in advertising campaigns, market shares are unmodified, $1/2$ each; (ii) the same happens if neither firm invests; while (iii) if one invests and the other does not, the former becomes a monopolist and the latter is thrown out of the market with zero demand. The game is represented in Matrix 4.5.

Now, if $F \in (0, \theta/2)$ – which is needed to ensure non-negative profits throughout – then Matrix 4.5 is an obvious prisoners' dilemma with (F, F) as the unique Nash equilibrium (in dominant strategies). In order to keep a positive market share, each firm invests in advertising, whereby the industry clearly overinvests as a result of individual incentives. Market shares are exactly the same as with no advertising at all, and the costs borne by the whole industry to set up and carry out the advertising campaign is entirely sunk.

Matrix 4.5 Advertising for market shares
in the Hotelling model

		2	
		0	F
1	0	$\theta/2; \theta/2$	$0; \theta - F$
	F	$\theta - F; 0$	$\theta/2 - F; \theta/2 - F$

4.1.4 Entry barriers

The R&D game describing an auction for a process or product innovation can also be interpreted as an entry game, where the incumbent may create a strategic entry barrier by winning the auction. Barriers to entry can be either natural or artificial, i.e., strategic. Natural barriers are given, for instance, by the initial investment that a firm has to undertake to enter a market, and clearly small newborn firms are disadvantaged compared to big ones. Artificial barriers may be build up strategically by incumbent firms in the absence of natural ones, to prevent further entries.

The analysis of entry barriers is a key aspect of industrial economics, dating back to Bain (1956) and Sylos-Labini (1962). Their approach relied on the possibility of an incumbent retaining monopoly power thanks to the so-called *Sylos postulate*, whose spirit can be summarized as follows. Examine a monopoly threatened by a potential entrant. The incumbent firm may announce that it will expand production to such an extent that it will be impossible for the entrant to obtain positive profits. The postulate states that, in the case of entry, the incumbent will indeed stick to the announcement and increase its output to annihilate the entrant. On this basis, the potential entrant will never materialize and the incumbent will remain a monopolist. From the late 1970s, game theory has been used to criticize the Sylos postulate and shed some new light on its weaknesses. To see how this works, we may examine a game due to Dixit (1979) and described by Matrix 4.6.

Players are an entrant, E, and an incumbent, I (the current monopolist). They are endowed with the same technology and sell substitute (but not necessarily identical) goods. E may choose whether to enter (e) or not to enter (ne), while I may either accommodate entry (a) or fight (f). The outcome (e, a) describes a duopoly with some positive profits equal to π_M for both firms (for instance, thanks to some degree of product differentiation). In (e, f) we observe a price war triggered by the incumbent and provoking symmetric losses (amounting, for instance, to a portion γrk, with γ positive but lower than one, of the fixed costs that both firms have to bear in order to set up their respective plants). Finally, if E does not enter, the industry remains in a monopoly regime and the distinction between a and f becomes immaterial to the incumbent as there is no strategic interaction at all.

This game has two Nash equilibria in pure strategies, (e, a) and (ne, f), but only the first is credible, since the second relies on the adoption of strategy c that is weakly dominated for the incumbent. Indeed, any preliminary announcement by I to the effect that the incumbent will fight entry is not credible. Accordingly, E will enter and a duopoly will obtain.

Matrix 4.6 The entry game

		I	
		a	f
E	e	$\pi_d; \pi_d$	$-\gamma rk; -\gamma rk$
	ne	$0; \pi_M$	$0; \pi_M$

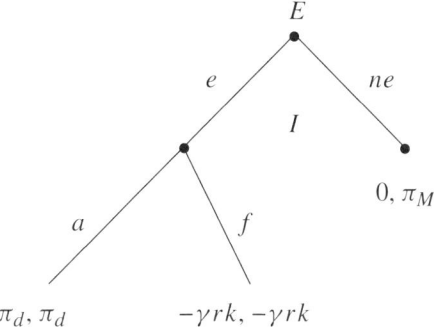

Figure 4.2 The entry game with sequential moves

Facing the *fait accompli*, the incumbent will rationally have no interest in sticking to the initial announcement. The outcome (e, a), besides qualifying as a Nash equilibrium, is also in weakly dominant strategies and, if the game is played sequentially (i.e., under perfect information) with E moving first, it is also subgame perfect: given that E correctly anticipates the incumbent's best reply, I's bluff is easy to unveil. The analysis of sequential play can be done by looking at the tree in Figure 4.2.

The truncated subgame on the right-hand side indicates that the absence of entry eliminates strategic interaction. Hence, one has to consider the left subgame only. This materializes only after entry, in which case the incumbent prefers a profitable duopoly to a harmful war. Anticipating this fact, the outsider will indeed enter, and drive the industry towards the subgame perfect duopoly equilibrium, irrespective of whether there has been an initial announcement by the incumbent to the effect that entry would meet opposition. Fighting – or announcing to be willing to fight – is *not credible*, as it is not part of the subgame perfect equilibrium. The bottom line of this game is that announcing something that is subgame perfect equilibrium is not necessary, as the rival operates under full information; otherwise, announcing a non-credible behaviour is just hot air and therefore it's a waste of time.[8] The following remark sums this up in a nutshell.

REMARK 4.4 In one-shot games, one shall not expect to observe artificial entry barriers based on aggressive pricing.

This is not the last word on entry barriers, though: we will revisit this game in the next chapter, where a repeated entry game based on Matrix 4.6 will reveal that the incumbent may find it convenient to fight in order to remain a monopolist. That is, time heals wounds – and also allows an incumbent firm to recover losses.

4.1.5 Networks, standardization and compatibility

An externality is an effect of individual behaviour that agents do not take into account when deciding a line of action. In the case of firms, an externality is a consequence of firms' strategies that is not reflected in prices and profits. The environmental effects and the depletion of natural resources caused by industrial activities and consumption are a typical example. We will come to these issues in the next section. The present section is for another phenomenon that you are probably acquainted with although you may never have interpreted it in the way industrial economists do. Nowadays, many markets and products are characterized by something called *network externality*. This label refers to the fact that people buy products or use services, producing both an intrinsic utility and an additional utility that increases with the number of people purchasing the same good or service. For instance, the laptop I'm using to write this book has an intrinsic utility as a typewriter, but also yields a positive network externality, as it allows me to connect to the web and do some Googling or to email my friends all over the planet.[9] Instead, my office desk has exclusively an intrinsic utility. Interestingly, there are goods that yield no intrinsic utility, but exclusively a network effect. An old-style phone would be useless to its owner, if the latter were the only person to own a telephone.[10]

The computer industry is perhaps the best example of a market where large-scale network externalities operate. And they are not the only relevant feature of the environment in which PCs, laptops and notebooks are connected through the web: *standardization* and *compatibility* also play a major role. Both aspects have much to do with hardware technology and the features of software – a sophisticated but intuitively clear form of product differentiation. As everybody knows, the computer industry is characterized by the presence of Apple with its own specific software, on one side, and IBM-compatible machines with Microsoft software, on the other.[11] They are neither standardized nor fully compatible, although the situation has improved a lot over the decades. I've never been an Apple user, at least not on a regular basis, because I've always studied and worked in places endowed with IBM-compatible stuff. Some 20 years ago, it would have been impossible or very difficult for me to read a document written using an Apple machine – and conversely for an Apple user trying to decode or convert a document produced on IBM-compatible hardware. Nowadays, it can be done in both directions with reasonable if not complete confidence. These considerations raise two related questions: (i) Why is the computer industry not using the same standard? (ii) Would it be privately (i.e., for firms) and socially optimal to adopt the same standard? The answer to the first question is that the network of people using IBM-like hardware reached a critical mass before the rival network did. Network externalities work through a bandwagon effect, whereby any new user decides which network to join on the basis of relative sizes, as the additional utility generated by the network is directly proportional to the number of people connected to it. The second question can be addressed using a couple of simple games. Matrix 4.7 describes the non-cooperative interplay between two firms, 1 and 2, using possibly different standards, *A* and *B*, with *B* being characterized by a higher intrinsic quality as compared to *A*. Information is complete, symmetric and imperfect. The problem is that firms do not internalize the externality – they only internalize profits, the latter being affected by product differentiation. Hence, the game presents two Nash equilibria along the secondary diagonal.

The consumers' standpoint is illustrated in Matrix 4.8, which exhibits the typical features of a pure coordination game with two Pareto-rankable Nash equilibria along the main

Matrix 4.7 The compatibility game

		2	
		A	*B*
1	*A*	10; 10	40; 80
	B	80; 40	30; 30

Matrix 4.8 Consumers' view
on compatibility

		2	
		A	*B*
1	*A*	20; 20	10; 15
	B	15; 10	70; 70

diagonal. The standardization of the industry is always preferable to the lack of standardization, and of course standardization on *B* is socially efficient. But this is not in the hands of consumers, alas.

The same problem can be investigated in a slightly more sophisticated model, where the quality differential separating the two standards appears explicitly. For the sake of simplicity, let's suppose that the population of consumers is partitioned into two groups, one gathering type-*H* consumers, the other gathering type-*L* consumers. The size of group *H* is M_H, while that of group *L* is M_L, with $M_H + M_L = 1$. The market is being supplied by two firms, *H* and *L*, each selling a standard of a given quality level, $\theta_H \geq \theta_L$. As in the Hotelling game, full market coverage prevails, so that each consumer buys one unit of the product, from one of the two firms. The two standards are sold at prices $p_H \geq p_L$. Market shares are, respectively, q_H and q_L, with $q_H + q_L = 1$. That is, everybody buys a unit, but in principle a type-*H* consumer may happen to patronize the low-quality standard (and conversely), and therefore the equalities $q_H = M_H$ and $q_L = M_L$ do not necessarily hold. This will depend upon quality and price differentials and the network externality. At this point, a simple and useful definition can be introduced.

DEFINITION 4.1 The industry is standardized on $i = H, L$ if all consumers patronize firm i, i.e., if $q_i = M_H + M_L$ and therefore $q_j = 0$, $j \neq i$.

In any other case, we may simply say that the industry is not standardized. Observe that standardization (or the lack thereof) is essentially determined by the distribution of demand. Accordingly, a crucial step is to describe consumer preferences in detail.

Define by U_{ij} the net utility of a type-*i* consumer buying a unit of the product whose quality standard is *j*. Indices *i* and *j* may or may not coincide. Also assume that the network externality is group-specific: sticking to the computer industry example, this amounts to saying that any Apple (IBM) user communicates with other Apple (IBM) users, but not to IBM (Apple) users. If a type-*H* consumer buys standard *H*, the resulting net utility is

$$U_{HH} = \theta_H - p_H + q_H \tag{4.10}$$

where quantity q_H measures the network externality, whose maximum size is equal to one when the market is standardized. If instead the same consumer chooses the low-quality alternative, net utility amounts to

$$U_{HL} = \theta_L - p_L + q_L - \delta \tag{4.11}$$

where parameter δ represents a disutility caused by the fact that our consumer belongs to group *H* but, for some reason, is buying standard L.[12] Since this disutility is in fact equivalent to a reduction in the network effect, a legitimate assumption is $\delta \in [0, 1]$. Similarly, a type-*L* consumer faces the following alternatives:

$$U_{LL} = \theta_L - p_L + q_L \tag{4.12}$$

$$U_{LH} = \theta_H - p_H + q_H - \delta \tag{4.13}$$

Now, everybody prefers standard *H* if the following condition holds:

$$\theta_H - p_H + 1 - \delta > \theta_L - p_L \tag{4.14}$$

The left-hand side of (4.14) measures the utility that a type-*L* consumer gets from buying standard *H* when everybody else does likewise: the network externality is maximized but diminished by the amount δ. The right-hand side measures the utility accruing to the same consumer as a result of buying standard *L* when he/she is the only consumer to patronize the inferior standard: in such a case there is no network externality at all, but the consumer doesn't have to bear the disutility δ. If (4.14) is satisfied, then the unilateral deviation from standard *H* to standard *L* is not convenient for a type-*L* consumer,[13] and therefore, *a fortiori*, it isn't either for a type-*H* consumer. Consequently, all consumers purchase the high-quality standard. It is worth noting that, in the limit case in which the two standards are sold at the same unit price ($p_H = p_L$), the above inequality would simplify to $1 - \delta > \theta_L - \theta_H$, which is surely true for all admissible levels of $\{\theta_H, \theta_L, \delta\}$.

Similarly, in order for all consumers to prefer standard *L*, it must be true that a unilateral deviation from *L* to *H* does not look appealing to a type-*H* consumer:

$$\theta_L - p_L + 1 - \delta > \theta_H - p_H \tag{4.15}$$

Conditions (4.14) and (4.15) can be appropriately rewritten to show the degree of differentiation $\Delta\theta = \theta_H - \theta_L$ on the one hand, and the price differential $\Delta p = p_H - p_L$ on the other. From (4.14) one obtains

$$\Delta\theta > \Delta p - 1 + \delta \tag{4.16}$$

while (4.15) yields

$$\Delta\theta < \Delta p + 1 - \delta \tag{4.17}$$

Given that

$$\Delta p + 1 - \delta > \Delta p - 1 + \delta \quad \text{for all } \delta \in [0, 1] \tag{4.18}$$

the discussion carried out so far implies the following result.

REMARK 4.5   If

$$\Delta\theta \in (\Delta p - 1 + \delta, \Delta p + 1 - \delta)$$

the industry may standardize on *H* or *L*, alternatively.

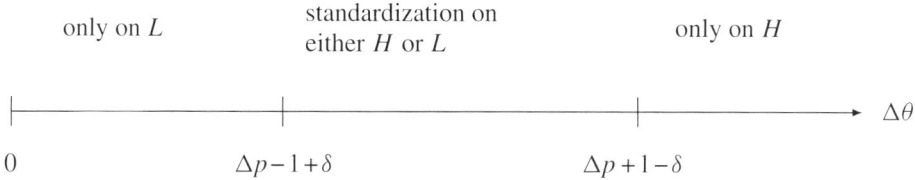

Figure 4.3 Quality and standardization

That is to say, if the quality differential takes intermediate values, then the market may sustain a single standard, but not necessarily the high-quality one. The above result can be translated into a more intuitive picture, as in Figure 4.3.

The intermediate region corresponds to the outcome spelled out in Remark 4.4. The left interval is one where only standardization on L is possible, due to a low degree of quality differentiation. Finally, in the right interval, quality differentiation is so high that the industry may standardize only on H.

4.2 Monetary and fiscal policies

So far we have dwelled upon games in industrial economics, a sub-field of microeconomics. However, game theory has been extensively applied to macroeconomics as well. An obvious area of interest, in this respect, is that of economic policy. Essentially, the latter has a twofold nature: monetary policy is the instrument for stabilizing prices (inflation), while fiscal policy is aimed at stabilizing aggregate demand (employment and income). The first is delegated to central bankers, while the second is manoeuvred by finance ministers, and ideally these policy-makers should (i) stick to their targets mandatorily, and (ii) coordinate their respective decisions to achieve Pareto-optimal outcomes for the economic system (say, a country or a group of countries, like the EU) as a whole.

From the point of view of the private sector (households, workers, etc.), monetary and fiscal policies exert their effects on employment, disposable income, and consumption patterns. Hence, in order to understand the nature and consequences of economic policy agendas, it is fair to set out with simple games between individuals taking monetary and fiscal stances as exogenously given. This approach to the problem explains the microeconomic foundations for the aggregate behaviour and performance of large economic systems.

4.2.1 Paying taxes

> The avoidance of taxes is the only intellectual
> pursuit that still carries any reward.
> > John Maynard Keynes

The following game has much in common with the very first one you saw at the beginning of this book, in Matrix 2.1. The version of the game you see in Matrix 4.9 has the very same structure, only slightly adapted so as to tell a story about the *private provision of public goods*. Players are two individuals or households, which may either pay (p) or evade (e) income taxes. The latter are to be invested in a public good (a hospital or a highway). The

Matrix 4.9 The private provision of public goods

		2	
		p	e
1	p	$100 + y - t; 100 + y - t$	$100 + y - t - \varepsilon; 100 + y$
	e	$100 + y; 100 + y - t - \varepsilon$	$y; y$

resulting amount of public good is independent of the amount of money raised through taxation, and, by definition of public good, the hospital or highway is accessible to both players alike, irrespective of whether both contributed to its construction or just one did. However, being the only contributor involves some disutility associated with the fact that one dislikes unilateral tax evasion. To fix ideas, let's say that gross income per capita is y, the amount of taxes is t per head, and the value that each player attaches to the public good is 100. It is reasonable to assume that $100 > y > t$. Disliking unilateral tax evasion brings about a decrease in utility equal to ε.

Since $100 + y - t < 100 + y$, there exists a strict incentive for each player to evade, provided the other player is paying taxes. Therefore, (p, p) cannot be a Nash equilibrium. This prompts the investigation of an asymmetric outcome: that is, can this game have a chicken-like structure? Suppose player 1 chooses e and player 2 chooses p. Is there an incentive for player 2 to unilaterally deviate from p to e? This depends on the relative size of y compared with $100 + y - t - \varepsilon$. In particular, if ε is large enough to yield $100 + y - t - \varepsilon < y$, or equivalently $100 - t - \varepsilon < 0$, then nobody pays taxes and the public good is not being produced, as a result of a prisoners' dilemma. Note that, for this outcome to arise as the only Nash equilibrium, each player has to be deeply disgusted by the tax evasion on the part of the other player, so much so that both of them end up preferring not to have the public good at all rather than offering a free lunch to others. Alternatively, if ε is small enough to yield $100 + y - t - \varepsilon > y$, Matrix 4.9 is a chicken game with two Nash equilibria in pure strategies along the secondary diagonal, and of course in this case the Nash equilibrium in mixed strategies also becomes relevant.

The above game offers a simple explanation of why people may be keen on evading taxes. The next step consists in looking at whether inspecting evasion turns out to be convenient or not.

4.2.2 *Fiscal inspection*

Given that people evade taxes with some positive probability, a government may introduce a fiscal inspection service whose explicit purpose is to establish how precise is the achievement of fiscal liabilities by each contributor. The effectiveness of inspection relies on some specific knowledge and the good professional qualification of the fiscal inspectors. The game illustrated in Matrix 4.10 involves an inspector, I and a taxpayer, T. The inspector may choose to inspect (i) or not (ni), the taxpayer may either pay (p) or evade taxes (e).

Matrix 4.10 The fiscal inspection game

		T	
		p	e
I	i	$-5; 0$	$5; -60$
	ni	$0; 0$	$0; 50$

Payoffs are set to tell the following story. Inspecting activities are costly, the cost of inspecting any single taxpayer being equal to 5. The amount of taxes to be paid is 100, and if the inspector detects evasion, the taxpayer has to pay the amount due plus a fine equal to 10. For simplicity, I am assuming that evasion is detected for sure. It is easily checked that this game has no Nash equilibrium in pure strategies: this, in a nutshell, explains why fiscal inspection is usually carried out on randomly drawn samples of taxpayers. Having reviewed the basics of individual behaviour, we may now turn to some macroeconomic policy games.

4.2.3   *Rules versus discretion*

> By a continuing process of inflation,
> government can confiscate, secretly
> and unobserved, an important part
> of the wealth of their citizens.
> > John Maynard Keynes

Whether a policy-maker should adopt an active – and possibly highly discretionary – behaviour or stick to well-defined and rigid rules has always been a hot issue in the debate on the optimal design of macroeconomic policies. Keynesians versus monetarists, neo-Keynesians versus neo-classicals, irrespective of the exact labels they have adopted over the decades, opposite schools have patronized one view or the other, alternatively. In general, a number of seemingly good reasons have been put forward to defend both rules and discretion. According to Milton Friedman (Nobel Prize in 1976) and the monetarist school, rules are preferable to activism, because a steady policy agenda reassures the economy and contributes to stabilize prices and the private sector's expectations, while Keynesians are more keen on fine-tuning practices. Two simple games, loosely based on Canzoneri (1985), help to illustrate the matter.

The game relies on the idea that an active macroeconomic policy may boost gross national product (GNP) but, in doing so, also creates inflation.[14] The government (G) may opt for either discretionary behaviour (d) or rules (r); the private sector may either gather information on the current inflation rate (i) or not (ni). In Matrix 4.11, information is costless. As a result, i is a dominant strategy for the private sector. In view of this, the government will find it convenient to go for rules rather than discretion, and (r, i) obtains as the unique Nash equilibrium, attainable by iterated deletion of dominated strategies. What if, more realistically, gathering information is a costly activity? The game becomes that depicted in Matrix 4.12.

The striking consequence is that the private sector no longer has a dominant strategy, and, as a result, the game yields no Nash equilibrium in pure strategies. Relatedly, there is no natural prescription for the government to choose either way. The foregoing discussion establishes the following remark.

Matrix 4.11 Rules versus discretion with costless information

		P	
		i	ni
G	d	−5; 5	10; −5
	r	5; 0	5; 0

Matrix 4.12 Rules versus discretion with costly information

		P	
		i	*ni*
G	*d*	−5; 4	10; −5
	r	5; −1	5; 0

REMARK 4.6   If the private sector is fully aware of the features of the economic environment and the consequences of economic policy, then rules looks preferable to discretion. If it is costly for the private sector to get the relevant information, then the adoption of discretionary behaviour by policy-makers cannot be ruled out *a priori*.

4.2.4   *Monetary and fiscal policy coordination*

Before abandoning the field of macroeconomic policy, I would like to show you a game illustrating in elementary but intuitive terms why central bankers (*CB*) and finance ministers (*FM*) should coordinate the menu of their respective policy agendas.

Each policy-maker may choose between restrictive (*r*) and expansionary (*e*) stances. From Matrix 4.13, we learn that *r* is dominant for the central bank, which controls inflation by lowering the money supply, while *e* is dominant for the government, which may boost employment and national income by increasing aggregate demand (via an increase in public expenditure or a decrease in fiscal pressure, or both). As a result, the game yields a unique Nash equilibrium, (*r*, *e*), at the intersection of dominant strategies. However, the game is a prisoners' dilemma as the other outcome along the secondary diagonal Pareto-dominates the Nash equilibrium. That is, policy-makers should indeed coordinate their respective strategies in order to achieve socially efficient outcomes.

4.3   Natural resources and the environment

To conclude this chapter, I will briefly discuss two related issues that would easily require a book of their own: the exploitation of natural resources; and the environmental effects of human production and consumption activities. Both have much to do with externalities and the undesirable tendency for firms not to internalize them at all. A further problem is that exhaustible natural resources are usually exploited to produce the energy required by polluting activities.

The exploitation of natural resources by a large number of agents (firms or individuals) gives rise to a gigantic prisoners' dilemma game that is traditionally known as the *tragedy of commons*, after the pioneering paper by Hardin (1968). The resource in question may either be renewable (e.g., fish) or non-renewable (e.g., fossil fuels). In the first case, an environmentally friendly approach would suggest the adoption of an exploitation rate strictly lower than the natural rate of reproduction of the resource itself. In the second case, any positive

Matrix 4.13 Macroeconomic policy coordination

		FM	
		r	*e*
CB	*r*	10; 2	5; 5
	e	8; 8	2; 10

Matrix 4.14 Tragedy of commons

$$
\begin{array}{c|cc}
 & \multicolumn{2}{c}{2} \\
 & l & h \\
\hline
1 \quad l & 10;\,10 & 4;\,20 \\
h & 20;\,4 & 12;\,12 \\
\end{array}
$$

exploitation rate will cause the resource to exhaust in a finite time, and therefore finding or inventing suitable alternatives not characterized by the same problem is or should be at the top of our agenda.

Let's see how a typical game of this sort works. Suppose it is a fishery game. Players are two firms, 1 and 2, whose respective harvest rates can be either high (h) or low (l), while the natural reproduction rate of fish is n, with $h > n > l$. Hence, harvesting at a low rate would be a wise strategy, but selfish strategic interaction unfortunately prevents the two firms from taking into account the depletion of the population of fish (whales could tell us a good deal about this). The resulting 2×2 game is shown in Matrix 4.14.

Each firm's payoff increases in its own harvest rate, given the rival's strategy. As a result, (h, h) is the unique Nash equilibrium, at the intersection of dominant strategies. This is the outcome players are bound to play, although they have a sense that preserving the natural resource for future exploitation would definitely be preferable.

An analogous situation arises if the use of the natural resource for production and/or consumption involves polluting the environment. Firms do not internalize externalities as, commonly, they only care about profits. And, in general, neither do single individuals: we all could easily switch off air conditioning or take buses or trains powered by electricity instead of driving our cars, but more often than not we don't, and we stick instead to our bad habits. Why? Because of a cheap excuse whereby we are 'atomistic' agents whose individual choices are, on aggregate, just irrelevant. Taken at face value, this view sounds very much like an intuitive and solid justification underlying the tragedy of commons, but of course its weakness can be quickly unveiled by trying to figure out what would be the effect of a major change in the attitude of a majority of people around the world.

Of course, a properly designed menu of policy measures may help, in this respect. For instance, taxes and subsidies can be used to induce firms to clean up their technologies or invest in R&D projects for new 'green' ones based on renewable energy sources. Two simple games illustrate this point. Matrix 4.15 describes an industry where two firms decide whether to invest in green technologies or not, in the absence of any regulation or policy measures.[15]

Strategies c and d stand, respectively, for *clean technology* and *dirty technology*. Inventing a green technology and making it a feasible one for large-scale production entails an R&D cost equal to k. The gross profits that each firm extracts from this market is π irrespective of the technology in use (say, because market shares and the equilibrium price are the same).

Matrix 4.15 Investing in green technologies, version 1

$$
\begin{array}{c|cc}
 & \multicolumn{2}{c}{2} \\
 & c & d \\
\hline
1 \quad c & \pi - k;\, \pi - k & \pi - k;\, \pi \\
d & \pi;\, \pi - k & \pi;\, \pi \\
\end{array}
$$

Matrix 4.16 Investing in green technologies, version 2

$$
\begin{array}{c|cc}
 & \multicolumn{2}{c}{2} \\
 & c & d \\
\hline
1 \quad c & \pi-k+\sigma\,;\pi-k+\sigma & \pi-k+\sigma\,;\pi-\tau \\
\quad d & \pi-\tau\,;\pi-k+\sigma & \pi-\tau\,;\pi-\tau \\
\end{array}
$$

Clearly, the game is a prisoners' dilemma whose only equilibrium (in dominant strategies) is (d, d).

Now suppose the government introduces an environmental policy whereby firms investing in green technologies are subsidized through public money by the amount $\sigma \in (0, k)$, while firms sticking to old and dirty technologies are taxed (or fined) by an amount $\tau \in (0, \pi)$. This regulated game looks as in Matrix 4.16.

If $\tau > k - \sigma$, strategy c is now the dominant one for both firms. Hence, if the subsidy is large enough, the unique equilibrium is (c, c). This outcome is also, by the way, Pareto-efficient.[16] Note that a similar outcome could be generated by a stimulus transmitted from environmentally concerned consumers telling firms that they would like to buy goods whose environmental impact is low or ideally nil, with no need to use public money for the same purpose. Summing up, we can express this in the following remark.

REMARK 4.7   In general, unregulated firms disregard the environmental implications of their strategies. The adoption of an appropriate tax/subsidy policy by the government, or the development of environmental awareness by consumers, can induce firms to internalize externalities and solve the tragedy of commons.

Further reading

For an exhaustive overview of the themes dealt with in this chapter, as well as many others belonging to industrial economics, see, at an increasing level of complexity, the textbooks by Cabral (2000), Shy (1995), Belleflamme and Peitz (2010), Tirole (1988) and Martin (2002). For specific topics, see: (i) Scotchmer (2004) on innovation; (ii) Beath and Katsoulacos (1991) and Anderson *et al.* (1992) for product differentiation; (iii) Shy (2001) on the economics of networks, compatibility and standardization; (iv) Blanchard (1997) and – at an advanced level – Blanchard and Fischer (1989), Persson and Tabellini (1990) and Barro and Sala-i-Martin (1995) on macroeconomic policy; and (v) Pearce and Turner (1989), Stern (2007, 2009), Tisdell (2009) and Anderson (2010) on environmental externalities and the exploitation of natural resources. The literature on auctions and mechanism design deserves special attention: see Vickrey (1962), Krishna (2002) and Klemperer (2003, 2004).

5 Repeated games and collusive behaviour

Time will say nothing but I told you so,
time only knows the price we have to pay;
if I could tell you I would let you know

W. H. Auden,
If I could tell you

So far, we have gone through a survey of one-shot games, where players meet just once and pursue the non-cooperative maximization of their respective objectives. In such games, calendar time (either past or future) has no bearing at all on agents' choices. Moreover, a striking feature of many games is that the resulting Nash equilibria may indeed be Pareto-inefficient (as typically happens in the prisoners' dilemma). However, casual observation suggests that players – firms, institutions, nations – often do interact repeatedly over long time spans. This prompts the construction of repeated game frameworks in which agents may collude to maximize common objectives, perhaps the sum of their individual ones, though keeping an intrinsically non-cooperative attitude. It is important to stress this latter aspect to distinguish the issue treated in this chapter from what belongs to the theory of cooperative games, which we shall deal with in Chapter 10. Essentially, the difference between repeated games and cooperative games can be grasped in the following terms. While players involved in a cooperative game are assumed to cooperate explicitly towards the maximization of a common objective irrespective of the time horizon (that is, even if the game is a one-shot event), a repeated game relies on the possibility that repeating over time the same constituent game gives rise to a form of *implicit* collusion based upon non-cooperative rules but nonetheless capable of replicating the outcome of an explicitly cooperative behaviour.

5.1 The prisoners' dilemma revisited

To fix ideas, let's recall the basic features of a prisoners' dilemma, by imposing the appropriate conditions on the payoff sequence characterizing a generic 2×2 game, as in Matrix 5.1.

Matrix 5.1 A generic prisoners' dilemma

		2	
		$\underline{s}$	$\tilde{s}$
1	$\underline{s}$	$\pi(\underline{s},\underline{s}); \pi(\underline{s},\underline{s})$	$\pi(\underline{s},\tilde{s}); \pi(\tilde{s},\underline{s})$
	$\tilde{s}$	$\pi(\tilde{s},\underline{s}); \pi(\underline{s},\tilde{s})$	$\pi(\tilde{s},\tilde{s}); \pi(\tilde{s}yy,\tilde{s})$

Strategy $\tilde{s}$ is strictly dominant for both players if and only if the following inequalities are simultaneously satisfied:

$$\pi(\tilde{s}, \underline{s}) > \pi(\underline{s}, \underline{s}), \quad \pi(\tilde{s}, \tilde{s}) > \pi(\underline{s}, \tilde{s}) \tag{5.1}$$

In such a case, $(\tilde{s}, \tilde{s})$ is the unique Nash equilibrium of this game, at the intersection of dominant strategies. Additionally, the game will be a prisoners' dilemma if

$$\pi(\tilde{s}, \tilde{s}) < \pi(\underline{s}, \underline{s}) \tag{5.2}$$

That is, players prefer the outcome $(\underline{s}, \underline{s})$, but they cannot reach it due to the presence of dominant strategies. Now we can ask ourselves whether there exists a mechanism whereby a prisoners' dilemma game may generate the Pareto-optimal outcome as an equilibrium via the repetition of the constituent game appearing in Matrix 5.1. The one-shot game, which may be labelled as $G(1)$, becomes the *constituent game* of the repeated game, or *supergame*, defined as $G(T)$ if the game is repeated over $t = 0, 1, 2, 3, \ldots, T$, or $G(\infty)$ if it is repeated infinitely many times, over $t = 0, 1, 2, 3, \ldots, \infty$. In either case, time is discrete.[1]

The concept underlying the results I am about to illustrate is that the perspective of receiving a long (possibly infinitely long) flow of future profits might change the players' perception of the game and the way they solve it, as compared to its one-shot version. As a preliminary step, I will briefly examine the role of the time horizon and time preferences through a simple example concerning something that all of you may have experienced in your everyday life.

5.2   Time and time discounting

> I never think of the future –
> it comes soon enough.
>> Albert Einstein

In the remainder, I will assume that players share the same time preferences, summarized by a common discount factor $\delta \in [0, 1]$. This is defined as $\delta \equiv 1/(1 + \rho)$, $\rho \geq 0$ being the (non-negative) discount rate applied to future payoffs. To fully appreciate the key role of time discounting, a little detour will be useful. The parameter ρ is, in fact, an interest rate used by agents in the repeated game so as to compare payoffs accruing to them in different periods of the supergame itself, in much the same way as one does (or should do) when evaluating investments in real or financial assets. To understand how this mechanism operates, consider the possibility of investing today a certain amount of money equal to M in government bonds (public debt) whose duration is a year. The bond yields an annual return rate equal to ρ, to be paid one year from now together with the full reimbursement of the initial sum. Therefore, after 12 months, you will get $M(1 + \rho)$, your net return being ρM. Now suppose you are fully satisfied with your investment and decide to reinvest the entire sum $M(1 + \rho)$ for yet another year in the same bond. At the end of the second year you'll get $M(1 + \rho)(1 + \rho) = M(1 + \rho)^2$. If you reinvest the initial sum plus the cumulated returns for t years, in the end you'll receive a sum equal to $M(1 + \rho)^t$.

Then imagine you run across a broker who proposes that you invest now the same amount M in some financial asset, and receive a sum $S > M$ after t years. Before signing the contract and buying that asset, you have to evaluate whether the (implicit) return rate is appealing or not. To do so, one thing you can do is to discount S at the rate being paid on government bonds of the same duration, t years. Hence, you compute the *present value*

$$PV = \frac{S}{(1+\rho)^t} = \frac{S}{\delta^t}$$ (5.3)

of the final payment, and compare it to the initial investment M. If $PV = M$, then the investment can be considered as fair, as it is equivalent to buying the same amount of public debt for the same number of years. Otherwise, if $PV > M$, it means that

$$S > M(1+\rho)^t = M\delta^t$$ (5.4)

i.e., this asset is more convenient than public debt,[2] and conversely if

$$PV < M \quad \Leftrightarrow \quad S < M(1+\rho)^t = M\delta^t$$ (5.5)

Finally, observe that PV becomes progressively smaller as the discount rate ρ increases (or equivalently, as the discount factor δ decreases): this fact will play a key role in what follows.

In order to have a preliminary appraisal of this mechanism and its implications, consider another perspective where you invest the sum M in a bond expiring after t years, and yielding a fixed coupon γ per year. At the end of the bond's life, the overall value of your investment, measured in present value, is equal to the discounted flow

$$M + \gamma + \frac{\gamma}{1+\rho} + \frac{\gamma}{(1+\rho)^2} + \frac{\gamma}{(1+\rho)^3} + \cdots$$ (5.6)

so that your net gain is the discounted flow of coupons, as the initial sum M is fully refunded.[3] Discounting takes place at the same rate ρ as above. If the bond is everlasting, it can be shown that[4]

$$\gamma + \frac{\gamma}{1+\rho} + \frac{\gamma}{(1+\rho)^2} + \frac{\gamma}{(1+\rho)^3} + \cdots = \gamma \sum_{t=0}^{\infty} \frac{1}{(1+\rho)^t} = \gamma \sum_{t=0}^{\infty} \delta^t$$ (5.7)

corresponds to

$$\gamma \times \frac{1}{1-\delta}$$ (5.8)

which is increasing with δ (or, equivalently, decreasing with ρ). Any increase in δ makes the value of future gains loom larger from the investor's standpoint.

Having settled these basic accounting issues, we may now turn our attention back to supergames.

5.3 Finite or infinite horizon?

> Time has been transformed, and we have changed;
> it has advanced and set us in motion;
> it has unveiled its face,
> inspiring us with bewilderment and exhilaration.
>
> Kahlil Gibran

Consider a repeated game whose constituent game is a two-player prisoners' dilemma, as in Matrix 5.1. The supergame is characterized by imperfect, complete and symmetric information in each stage (that is, at any time t), while information is perfect across any pair of contiguous stages, $(t-1, t)$ or $(t, t+1)$: this means that in each period both players are able to reconstruct the entire sequence of moves up to that period.

As to the length of the time horizon (i.e., the duration of the supergame), there are three different possibilities:

- the terminal date is finite, say, T, and both players are aware of this;
- the supergame lasts forever, the terminal date being ∞, or doomsday, and both players are aware of this; and
- the supergame ends at some finite date T, and both players know that it is finite but they have no precise idea about when the supergame will in fact end – in view of this, every day both players attach a positive probability $\mathfrak{p} \in (0, 1)$ to the fact that the supergame will indeed continue.

The third case seems a mix of the first two, while in fact it can be shown that it yields the same results as the second, except that the mathematics involved is more cumbersome. Therefore, in order to keep the analysis as easy as possible, I will confine myself to an intuitive comparative appraisal of the first two perspectives.

The fundamental question is this: Will a supergame with finite terminal date (*a priori* known to players) work? That is, will it allow players to generate a Pareto-efficient solution to the prisoners' dilemma, or not?

It can be quickly shown that the answer is negative. To see this, consider the finite supergame $G(T)$, with the stage game being a prisoners' dilemma at each $t = 0, 2, 3, \ldots, T$, and solve it by backward induction from T. Corresponding to the last stage taking place at the terminal time, there is no future by definition. Accordingly, players will solve the prisoners' dilemma by choosing their respective dominant strategies, as if they were interacting just once. This generates the Pareto-inefficient Nash equilibrium of the one-shot game we already know all too well. Now you may step back one period, to see what happens at $T - 1$. At this date, the fact that there exists a (short) future is clearly immaterial, as players know that the one-shot inefficient equilibrium will pop up at T. Hence, at $T - 1$ they will behave as if it were the terminal time, reproducing the one-shot prisoners' dilemma – as if the future did not exist. This mechanism reproduces itself time and again as players step back towards the initial date, at which any future has become immaterial. Therefore, they play the Nash equilibrium in dominant strategies of the constituent game $T + 1$ times, being altogether unable to construct a remedy to the prisoners' dilemma. This result can be summarized in the following remark.

REMARK 5.1   The only subgame perfect equilibrium of a prisoners' dilemma repeated over a finite horizon consists in repeating the Nash equilibrium in dominant strategies of the constituent game at every stage.

It is worth noting the quite strong analogy between this set-up and Rosenthal's centipede game: in both cases, backward induction, together with the presence of dominant strategies at each stage, generates a complete disaster that prevents players from achieving Pareto efficiency.[5] Consequently, we'll necessarily have to cope with supergames of infinite duration, which can be viewed as a simplified approach to more realistic problems where agents interact a finite number of times but the terminal date is, so to speak, 'foggy'.

5.4   The folk theorems

> Time present and time past
> are both perhaps present in time future,
> and time future contained in time past.
> If all time is eternally present,
> all time is unredeemable.
>
> T. S. Eliot,
> *Burnt Norton* (*Four Quartets*)

The theory of repeated games is essentially based on a genealogy of theorems whose origins have to be sought in the corridors of mathematics and economics department in the 1960s. Since the father is unknown – or, at least, there is no unanimous agreement on paternity[6] – these results fall under the commonly accepted label of *folk theorems*.

A seemingly obvious benchmark for understanding repeated games and the emergence of collusion consists in taking a look at cartels. Firms collude in prices or market shares (or both) to increase their profits as compared to fully non-cooperative equilibria. In this regard, one may think of OPEC (Organization of Petroleum Exporting Countries) as a well-known example, while actually it is not an appropriate one, at least in terms of what we are about to learn concerning supergames. Why? Because OPEC is an *explicit* cartel agreement among sovereign countries, instead of firms: as such, OPEC members cannot be prosecuted, while firms are subject to antitrust laws worldwide. This fundamental difference is the reason why OPEC members usually declare in public what they are about to do with the oil price almost on a daily basis, while colluding firms prefer to keep the lowest possible profile.

Cartel behaviour is of course harmful, as it decreases consumer surplus and social welfare. To understand how this happens, look at Figure 5.1, offering a graphical sketch of a market with a linear demand $p = a - Q$, where p is the unit price of the good being traded, Q is the overall quantity being traded, and $a \geq p$ is the so-called *reservation price* (the maximum price that any consumer in this market would be willing to pay for the first unit). Production takes place at a constant unit cost c, positive and lower than a.

This market is being served by a given number of firms (at least one), with a resulting quantity–price equilibrium pair identified by the point (Q^*, p^*) along the demand function. The area of the rectangle Π measures industry profits (or producer surplus), while the areas of the two triangles CS and DL measure consumer surplus and the deadweight loss, respectively. Consumer surplus is the amount of surplus that consumers would be willing to spend to purchase each and every unit of Q^*, but are not obliged to pay because every unit is sold at price p^*. The deadweight loss is the portion of surplus that vanishes because of firms' market power, every time that a market is not working under perfectly competitive conditions

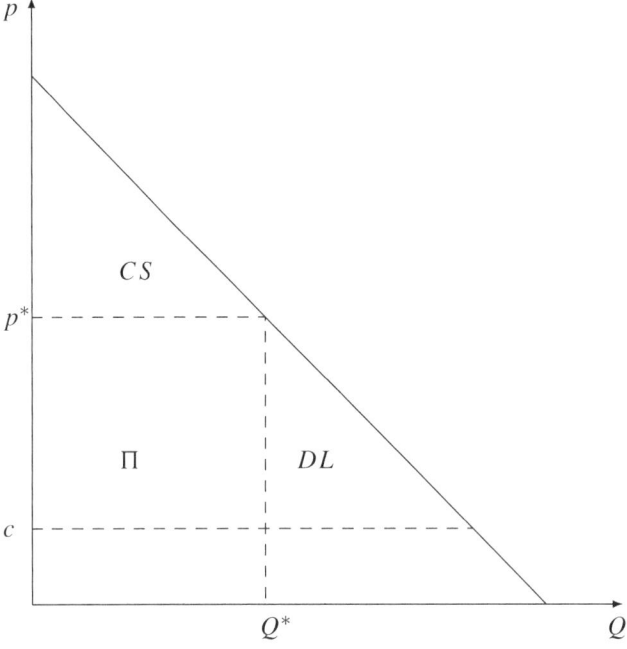

Figure 5.1 Profits, consumer surplus and welfare

(whereby firms should price at marginal cost). The sum of industry profits and consumer surplus, $SW = \Pi + CS$, is a commonly accepted measure of social welfare.[7] Any price increase, accompanied by an output reduction, enlarges the deadweight loss and ultimately decreases welfare.

Now recall that an outcome equivalent to perfect competition also arises under the (admittedly quite unrealistic) Bertrand paradox that we met in Chapter 4. Shall we think that, whenever price-setting firms selling a homogeneous good succeed in keeping the price above c, they are colluding? Not necessarily. A stylized version of the Cournot (1838) model suffices to make the point. To begin with, the Bertrand game replicates perfect competition provided that firms are not capacity-constrained, i.e., each of them has to be endowed with a plant whose capacity k is sufficiently large to serve the entire market at $p^{BN} = c$. As we already know, this is a rather strong assumption. What if, more reasonably, both firms choose their respective capacity before engaging in price competition – with the rational expectation to operate at full capacity corresponding to the resulting equilibrium? This is a brilliant intuition by which Kreps and Scheinkman (1983) have shown that price and quantity competition are the same under capacity constraint. Without delving into the mathematical details of their proof, I will give just an intuitive explanation of the basic result. Since building up capacity is costly – say, rk if the unit price of capital goods is r – one should not expect firms to acquire the amount of capital needed to drive the market price down to c, the reason being that this would entail *negative* profits, with firms being unable to cover the cost of capacity because gross profits would be nil due to the standard Bertrand argument. Therefore, firms will install plants whose overall size falls short of the perfectly competitive output of the industry, i.e., they will practice a form of demand rationing enabling them to keep the price above unit cost c. In practice, this approach is equivalent to saying that firms choose their respective output levels and then let the market clear through the price mechanism. The result is that they solve the Bertrand paradox. This observationally coincides with (and replicates the spirit of) the original quantity-setting game examined by Cournot (1838). Thus, any pre-commitment on output or capacity levels has an intrinsically anticompetitive flavour, but can well be the outcome of a fully non-cooperative behaviour and, as such, it may be prosecuted *per se*, unless one can show that some agreement intentionally aimed at raising prices is hidden behind the curtains.

Explicit agreements, however, are likely to be detectable. The purpose of this chapter is to illustrate other – implicit – mechanisms that can be used to generate and stabilize cartels over time, without yielding any hard evidence that could be used against cartel members in court. This is the specific aspect of firms' activities that has operated as an engine of the literature on repeated games and generated its folk theorems.

The most commonly used version – and the first to be published in an academic journal – is due to Friedman (1971), and I will stick to it throughout this chapter as well as in further applications in the remainder of the book.

To illustrate the theorem, I will use the Bertrand duopoly game as a benchmark case. As usual, firms 1 and 2 are fully symmetric and sell the same good. The unit production cost is c and any fixed costs are assumed away. As a consequence, the resulting Bertrand–Nash equilibrium profits are nil. By contrast, suppose firms build up a cartel to maximize joint profits, by setting the price at its monopoly level, and then split the resulting monopoly profits evenly. If so, individual cartel profits are $\pi_C = \pi_M/2$. Yet, the temptation to deviate from monopoly pricing is there: setting $p_D = p_M - \varepsilon$, with ε positive and arbitrarily small, the cheating firm throws the 'loyal' cartel mate out of the market and (disregarding the discount ε) gains full monopoly profits $\pi_D = \pi_M$. The resulting 2×2 game is represented in Matrix 5.2.

Since strategy p_D weakly dominates strategy p_M, and the resulting unique Nash equilibrium (p_D, p_D) is Pareto-inefficient (for firms, of course), the Bertrand game is a prisoners' dilemma.

The situation faced by firms is illustrated in Figure 5.2, which is drawn in the profits space. The origin of the axes denotes the Bertrand paradox, with profits equal to zero for both firms, while the line at $-45°$ is the frontier of industry profits, along which $\pi_1 + \pi_2 = \pi_M$.

With the exclusion of the origin, the triangular region identified by the intersections between the negatively sloped frontier and the axes defines the 'cone' containing all the outcomes that (at least weakly) Pareto-dominate the Bertrand paradox itself. If there were capacity constraints or some degree of product differentiation, the non-cooperative outcome would put firms at point N, and the region of Pareto-superior outcomes would be the triangle NBB'.

Any cartel sustaining prices above their fully non-cooperative level would drive the industry to some point in the relevant cone, which one may label as the *admissible region* for collusion. Full collusion at the monopoly price, as in Matrix 5.2, allows firms to reach the

Matrix 5.2 Monopoly versus Nash pricing in the Bertrand model

		2	
		p_M	p_D
1	p_M	$\pi_M/2$; $\pi_M/2$	0; π_M
	p_D	π_M; 0	0; 0

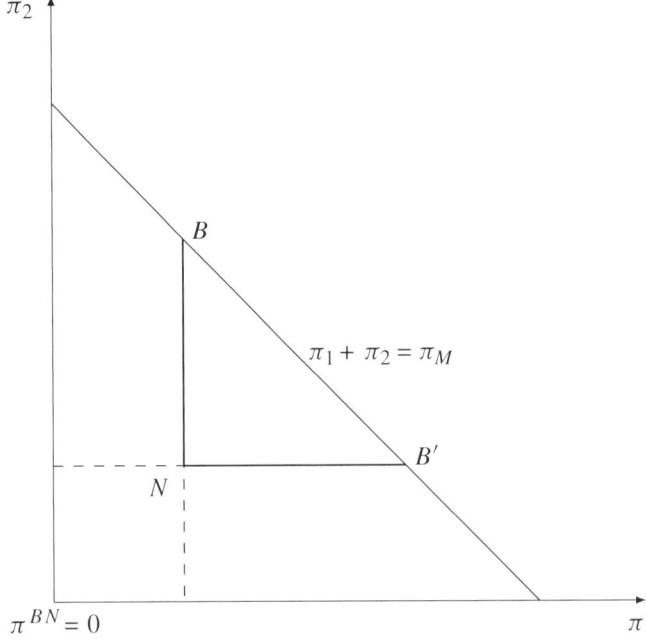

Figure 5.2 The frontier of industry profits and the cone of Pareto-superior allocations

frontier. Incidentally, this graphical representation of the cartel problem clearly shows that there exist infinitely many degrees of collusion, monopoly pricing being only one out of many. In particular, should firms be unable to stabilize full collusion, there remains other less attractive – but also less demanding – alternatives, something that clearly makes the antitrust officers' life a dismal one.

We are now in a position to ask ourselves how the equilibrium outcome of an infinitely repeated game whose constituent stage game is a prisoners' dilemma may be modified because of repetition.

The most general formulation of the folk theorem for an infinite supergame $G(\infty)$ is the following.

THEOREM 5.1   (FOLK THEOREM) *The supergame $G(\infty)$, during which players share the same constant intertemporal discount factor $\delta \in [0, 1]$, may yield as an equilibrium any outcome such that the resulting payoff vector belongs to the admissible region, provided that the discount factor δ is sufficiently high.*

Theorem 5.1 may appear, at first sight, a little bit cryptic. In fact, it simply says that a Pareto improvement over the Nash equilibrium of the constituent game $G(1)$ can be generated through an infinite repetition of the latter. How big this Pareto improvement will be, well, that depends on how patient (or forward-looking) the players are. That is, players can attain any payoff vector in the region NBB', provided they attach a sufficiently high weight to future periods. The only case where this achievement is ruled out altogether is that where agents are totally short-sighted, i.e., $\delta = 0$. If so, they perceive each stage of the supergame as if it were the only stage, and consequently are forced to play the Nash equilibrium of the one-shot game forever.

To give you the flavour of how a supergame operates, I will illustrate the infinite repetition of Matrices 5.1 and 5.2 (i.e., a generic prisoners' dilemma and the Bertrand duopoly game) using the so-called *perfect folk theorem* due to James Friedman (1971). Then, I will outline an alternative view on repeated games using Axelrod's (1981) tit-for-tat strategies, widely used in political science.

5.4.1   *The perfect folk theorem*

> If people are good only because they fear punishment,
> and hope for reward, then we are a sorry lot indeed.
>
> Albert Einstein

Friedman's version of the folk theorem is called *perfect* because it produces a subgame perfect equilibrium. To begin with, one has to spell out the rules governing the behaviour of players during the supergame. Players receive the following instructions, which they have to abide by throughout the repeated game:

1   In period 0, play the collusive strategy (for instance, monopoly price p_M).
2   In any period $t \geq 1$, play the collusive strategy, provided everybody has adopted it at time $t - 1$. Otherwise, play the one-shot Nash equilibrium strategy (for instance, the Bertrand–Nash price).

These strategies are known as *grim trigger strategies* or just *trigger strategies*, as any deviation from the collusive path triggers a Nash punishment that will last forever – with no chance of getting back to collusion ever again – due to the recursive nature of the second rule. This Nash reversion, or Nash punishment, is meant to exert a deterrence effect against unilateral deviations from the cartel path.

To see how these rules may succeed in stabilizing collusion forever, let's focus first on the repetition of the game described in Matrix 5.1 under the conditions spelled out in conditions (5.1) and (5.2). The collusive outcome is $(\underline{s}, \underline{s})$. If both players stick to it forever, each of them receives the following flow of discounted payoffs:

$$\Pi_C = \pi(\underline{s}, \underline{s}) + \delta\pi(\underline{s}, \underline{s}) + \delta^2\pi(\underline{s}, \underline{s}) + \delta^3\pi(\underline{s}, \underline{s}) + \cdots$$
$$= \pi(\underline{s}, \underline{s})(1 + \delta + \delta^2 + \delta^3 + \cdots) \tag{5.9}$$

which, using the techniques illustrated above, simplifies to

$$\Pi_C = \frac{1}{1 - \delta} \times \pi(\underline{s}, \underline{s}) \tag{5.10}$$

This payoff stream measures the individual firm's incentive to remain on the collusive path forever. Yet, given that the stage game is a prisoners' dilemma, the temptation to defect is always there, at any time t during the supergame. Moreover, if deviation is attractive, a cheating player will indeed find it most attractive at time zero, as this means that the resulting payoff is not diminished by time discounting.[8] However, defection is punished *à la*

Nash forever. Hence, the flow of payoffs associated to the alternative path starting with a unilateral deviation is

$$\Pi_D = \pi(\tilde{s}, \underline{s}) + \delta \pi(\tilde{s}, \tilde{s}) + \delta^2 \pi(\tilde{s}, \tilde{s}) + \delta^3 \pi(\tilde{s}, \tilde{s}) + \cdots \tag{5.11}$$

Now observe that

$$\delta \pi(\tilde{s}, \tilde{s}) + \delta^2 \pi(\tilde{s}, \tilde{s}) + \delta^3 \pi(\tilde{s}, \tilde{s}) + \cdots = \frac{1}{1-\delta} \times \pi(\tilde{s}, \tilde{s}) - \pi(\tilde{s}, \tilde{s})$$

$$= \left(\frac{1}{1-\delta} - 1 \right) \pi(\tilde{s}, \tilde{s}) \tag{5.12}$$

and therefore

$$\Pi_D = \pi(\tilde{s}, \underline{s}) + \frac{\delta}{1-\delta} \times \pi(\tilde{s}, \tilde{s}) \tag{5.13}$$

Players will find it convenient to collude forever if and only if the following inequality holds:

$$\Pi_C \geq \Pi_D \quad \Leftrightarrow \quad \frac{\pi(\underline{s}, \underline{s})}{1-\delta} \geq \pi(\tilde{s}, \underline{s}) + \frac{\delta \pi(\tilde{s}, \tilde{s})}{1-\delta} \tag{5.14}$$

which can be rearranged to yield Friedman's stability condition:

$$\delta \geq \frac{\pi(\tilde{s}, \underline{s}) - \pi(\underline{s}, \underline{s})}{\pi(\tilde{s}, \underline{s}) - \pi(\tilde{s}, \tilde{s})} = \delta^* \tag{5.15}$$

The fundamental result summarized in (5.15) deserves a closer look. The fraction on the right-hand side is always less than one because $\pi(\underline{s}, \underline{s}) > \pi(\tilde{s}, \tilde{s})$, as the underlying stage game is a prisoners' dilemma. Hence, (5.15) identifies a threshold of the discount factor, δ^*, above which collusion is stable forever. This indeed captures the idea that, in order to solve a prisoners' dilemma once and for all, players have to be far-sighted enough. This is a general property of folk theorems, differing in a number of details, but consistently generating the same essential message: if future payoffs loom large enough from the players' standpoint, then an infinitely repeated game yields a way out of the inefficiency affecting the one-shot version of the same game.

Additionally, observe that the threat of reverting to the infinite punishment after any deviation is indeed a credible one, as the punishment itself is an infinite chain of Nash equilibria. Therefore, Friedman's rules produce a subgame perfect equilibrium.

Let's now turn to the Bertrand duopoly game with fully symmetric firms. For simplicity, we can examine the case of full collusion, with firms examining the possibility of setting the monopoly price over $t = 0, 1, 2, 3, \ldots, \infty$.

The relevant payoffs appear in Matrix 5.2. An everlasting cartel yields to each member the following discounted profit flow:

$$\Pi_C = \frac{1}{1-\delta} \times \frac{\pi_M}{2} \tag{5.16}$$

According to rule 2 above, a unilateral deviation gives the cheating firm full monopoly profits for a single period only, as firms shall subsequently revert to the Nash punishment forever – which corresponds to an infinite replication of the Bertrand paradox annihilating profits. Accordingly, we have that

$$\Pi_D = \pi_M + 0 \tag{5.17}$$

The stability condition requires

$$\frac{\pi_M}{2(1-\delta)} \geq \pi_M \quad \Leftrightarrow \quad \delta \geq \frac{1}{2} = \delta_B^* \tag{5.18}$$

This result is an alarm bell ringing loud on the desks of any antitrust agency. Basically, it says that these two firms are very likely to collude, since

$$\delta_B^* = \frac{1}{1+\rho_B^*} = \frac{1}{2} \tag{5.19}$$

and thus (5.18) can be rewritten in a more explicit way as $\rho \leq \rho_B^*$, with $\rho_B^* = 1$ and therefore $\rho \leq 1$. This latter inequality says that using any interest rate lower than 100 per cent to discount future profits will indeed allow firms to stabilize the price cartel forever. If entrepreneurs evaluate the cartel in the same way as a long-term investment project, and use long-term interest rates to discount profits in the repeated game, then obviously this cartel should be stable, as interest rates on long-run investments are much lower than 100 per cent.

Admittedly, this is a rather extreme case, as product differentiation, cost asymmetries, etc., are assumed away. Moreover, cartels may involve more than two firms, and it can be easily shown that cartel size negatively affects the stability of collusion. To appreciate this scale effect, we may stick to the initial assumption of price-setting behaviour with fully symmetric firms and a homogeneous good, except that we consider a population of n cartel members.[9]

The firms aim at stabilizing full collusion on the frontier of monopoly profits over an infinite horizon. Hence, along the collusive path, each of them gets a symmetric share of monopoly profits π_M/n per period. As in the duopoly case, a unilateral deviation yields full monopoly profits to the deviator and triggers an infinite punishment via the repetition of the Bertrand–Nash equilibrium forever. Therefore, increasing the size of the cartel rescales the profitability of collusion downwards, while the appeal of a deviation and the harshness of the punishment remain exactly the same as in the duopoly case. Condition (5.18) becomes

$$\frac{\pi_M}{n(1-\delta)} \geq \pi_M \quad \Leftrightarrow \quad \delta \geq \frac{\pi_M - \pi_M/n}{\pi_M} = \frac{n-1}{n} \tag{5.20}$$

where the critical threshold of the discount factor is monotonically increasing with the number of firms participating in the cartel. In the limit, with n infinitely large, the fraction $(n-1)/n$ tends to one, showing that an infinitely large cartel is unsustainable. The intuitive reason for this result is that, the bigger the cartel, the smaller the slice of monopoly profits accruing to any cartel member. In the limit, the size of the slice tends to zero and therefore the difference between the performance of a cartel and that of a perfectly competitive market tends to vanish. This result is a robust one (that is, it is confirmed by fully fledged models admitting the presence of some degree of product differentiation and other realistic features) and conveys the basic message that small cartels are much more likely to arise and persist than large ones.

As I have already said, the folk theorem has evolved over time (and keeps doing so) to deal with several important aspects of repeated games. A very relevant issue is the efficiency of punishment. In this regard, *optimal* (or, most efficient) *punishments* have been designed by Abreu (1986, 1988) and Fudenberg and Maskin (1986). Other issues are imperfect monitoring of deviations (see Green and Porter, 1984), the possibility of observing countercyclical collusion, with firms turning to non-cooperative behaviour during economic booms (see Rotemberg and Saloner, 1986), the influence of product differentiation on cartel stability (see Deneckere, 1983; Chang, 1991, 1992; Ross, 1992, *inter alia*) and the choice of productive capacity in oligopoly supergames (see, e.g., Brock and Scheinkman, 1985).

5.4.2 Tit-for-tat

The structure of the supergame appearing in Axelrod (1981) has much in common with Friedman's perfect folk theorem, while some of the features are specific to the *tit-for-tat* logic and do not appear anywhere else in the literature on implicit collusion in repeated games.

Also in Axelrod's model, the supergame takes place over an infinite horizon, the constituent game being a prisoners' dilemma (i.e., conditions (5.1) and (5.2) are assumed to hold). What changes is the set of rules regulating the behaviour of agents throughout the supergame:

1   In period 0, play the collusive strategy.
2   In any period $t \geq 1$, play the collusive strategy, provided everybody has adopted it at time $t - 1$. Otherwise, play the strategy chosen by your rival(s) in period $t - 1$.

Hence, the second rule does not correspond to rule 2 in Friedman (1971). The label *tit-for-tat* expresses this idea of reciprocating the rival's behaviour. To illustrate how Axelrod's rules operate, I will refer to Matrix 5.1. In a nutshell, the tit-for-tat rule establishes that a player that never colludes will receive $\pi_i(\tilde{s}, \underline{s})$ at $t = 0$ and $\pi_i(\tilde{s}, \tilde{s})$ at any $t \geq 1$. Applying Axelrod's rules, one replicates condition (5.14), which obviously implies that collusion is stable provided (5.15) is met. As in Friedman (1971), also here stability is a matter of time discounting. However, in the tit-for-tat framework this is not the only possibility. Following Axelrod (1981, p. 311, Theorem 2), one can verify that adopting the tit-for-tat strategy is more convenient than alternating strategies $\underline{s}$ and $\tilde{s}$ if

$$\frac{\pi_i(\tilde{s}, \underline{s}) + \delta \pi_i(\underline{s}, \tilde{s})}{1 - \delta^2} \leq \frac{\pi_i(\underline{s}, \underline{s})}{1 - \delta} \tag{5.21}$$

that is, if

$$\delta \geq \frac{\pi_i(\tilde{s}, \underline{s}) - \pi_i(\underline{s}, \underline{s})}{\pi_i(\underline{s}, \underline{s}) - \pi_i(\underline{s}, \tilde{s})} = \hat{\delta} \tag{5.22}$$

Accordingly, a player will indeed adopt tit-for-tat strategies in equilibrium provided that

$$\delta \geq \max \left(\frac{\pi_i(\tilde{s}, \underline{s}) - \pi_i(\underline{s}, \underline{s})}{\pi_i(\tilde{s}, \underline{s}) - \pi_i(\tilde{s}, \tilde{s})}, \frac{\pi_i(\tilde{s}, \underline{s}) - \pi_i(\underline{s}, \underline{s})}{\pi_i(\underline{s}, \underline{s}) - \pi_i(\underline{s}, \tilde{s})} \right) \tag{5.23}$$

Observe that the two expressions appearing in the parentheses on the right-hand side of inequality (5.23) share the same numerator, so that the difference between the two critical thresholds of the discount factor depends entirely on their respective denominators. Provided conditions (5.1) and (5.2) apply – so as to make the constituent game a prisoners' dilemma – the ranking of δ^* and $\hat{\delta}$ will be sensitive to the relative size of payoffs.

5.4.3 *Examples*

The present chapter is probably the most demanding of the entire book. For illustrative purposes, I will dwell upon two simple games where payoffs take numerical values.

Consider the following 2×2 games (see Matrices 5.3 and 5.4), both sharing the structure of the prisoners' dilemma.

Suppose the players interact over an infinite horizon, sharing the same discount structure represented by a common discount factor δ. You may compute the threshold values δ^* and $\hat{\delta}$ for both supergames, and easily rank them.

What you get on the basis of Matrix 5.3 is

$$\delta^* = \frac{5-3}{5-1} = \frac{1}{2}, \quad \hat{\delta} = \frac{5-3}{3-0} = \frac{2}{3} \tag{5.24}$$

so that $\delta^* < \hat{\delta}$. Instead, from Matrix 5.4, you have that

$$\delta^* = \frac{4-3}{4-2} = \frac{1}{2}, \quad \hat{\delta} = \frac{4-3}{3-0} = \frac{1}{3} \tag{5.25}$$

and therefore $\delta^* > \hat{\delta}$. However, it is worth noting that the difference between trigger strategies and tit-for-tat ones is not a mere issue of relative stability as might seem to be the case from (5.24) and (5.25): indeed, a tit-for-tat profile is, in general, unable to yield subgame perfect equilibria as shown by Kalai *et al.* (1988).[10]

Matrix 5.3

		2 *l*	*r*
1	*t*	3; 3	0; 5
	b	5; 0	1; 1

Matrix 5.4

		2 *l*	*r*
1	*t*	3; 3	0; 4
	b	4; 0	2; 2

5.4.4   *Experimental evidence*

Starting with Axelrod (1981, p. 309), the bottom line of repeated games – namely, that repetition may endogenously solve the prisoners' dilemma, inducing players to generate Pareto-efficient equilibrium outcomes – has been tested in several controlled experiments, which have singled out the following aspects.

1   The *shadow of the future* affects players' behaviour in a way that appears to be closely in line with theoretical predictions (see Dal Bo, 2005).
2   Yet, the matching procedure (be it random or not) among players involved in the experiments is indeed important, as it emerges that agents that have never met before tend to deviate more frequently than others that are already acquainted with each other (Yang *et al.*, 2007). In this respect, though, it must be said that there is no unanimity, as some recent experiments show that cooperation can be sustained even in anonymous settings (see, e.g., Camera and Casari, 2009).
3   If the payoff associated with reciprocal defection (i.e., with the Nash equilibrium of the one-shot game) is negative, then players choose the dominated strategy – and therefore collude – more frequently than otherwise. This seems to reveal the presence of a strong risk aversion in players belonging to the samples involved in several experiments that have exhibited this type of regularity (Sabater-Grande and Georgantzis, 2002 and Hauk, 2003).

5.5   The chain store paradox

The theory of repeated games has been used to work out the conditions for implicit collusion, but it is also a powerful instrument to probe the robustness of other results, generated by games other than the prisoners' dilemma. In Chapter 4, we have encountered the issue of strategic entry barriers in a one-shot game illustrated by Matrix 4.6. Here, we will revisit the same problem in a more general framework where the entry game repeats over time: this is the famous *chain store paradox* proposed by Selten (1978).

The structure of the constituent game is essentially the same as Matrix 4.6. Players are an incumbent, I, and an entrant, E. The entrant may either enter (e) or not (ne); the incumbent may either accommodate (a) or fight (f) entry. The resulting Matrix 5.5 is slightly more general than Matrix 4.6, with the payoff sequence

$$\pi_{\mathrm{M}} > \pi_{\mathrm{d}} > 0 > \pi_{\mathrm{w}} \tag{5.26}$$

with the negative payoff π_{w} denoting the symmetric loss that firms will incur if, say, the incumbent triggers a price war after entry has taken place.

As we already know, the one-shot version of this game has two pure-strategy Nash equilibria: (e, a), where a duopoly obtains; and (ne, f), where the incumbent remains a monopolist. However, given the chain of inequalities in (5.26), the threat of adopting an aggressive

Matrix 5.5 Selten's entry game

		I	
		a	f
E	e	$\pi_{\mathrm{d}}; \pi_{\mathrm{d}}$	$\pi_{\mathrm{w}}; \pi_{\mathrm{w}}$
	ne	$0; \pi_{\mathrm{M}}$	$0; \pi_{\mathrm{M}}$

behaviour is not credible, as f is a weakly dominated strategy for the incumbent. Conversely, the outcome (e, a) is in weakly dominant strategies and is also subgame perfect in the sequential play version of the game, with the entrant moving first. The proof of this result is omitted for brevity as the game tree replicates Figure 4.2.

So far, no big news. The novelty of Selten's approach to the analysis of strategic entry barriers consists in admitting the realistic possibility that the entry game be repeated, either over time or on several markets. That is, Selten builds up a supergame that summarizes in a single set-up two alternative but formally equivalent situations:

- The incumbent owns a single outlet and faces a repeated entry threat by either the same potential entrant or a population of symmetric ones, in a single market.
- The incumbent owns many identical outlets in as many identical local markets, and faces a threat of entry by either the same potential entrant or a population of symmetric firms, one in each single market.

The second interpretation of the supergame is responsible for the label that Selten chose for his model, known as the *chain store paradox*.

In this set-up, calendar time and space – or the number of markets – are formally equivalent and can be interchanged with one another without affecting the result. Therefore, in line with the spirit of this chapter, I will treat the problem as a repeated game and refer to time instead of space. Having settled this issue, the problem is then whether the game repeats over a finite or infinite horizon. On the basis of Remark 5.1, and without dwelling upon the details of the proof, I may claim the following remark.

REMARK 5.2 If the entry game is repeated over $t = 0, 1, 2, 3 \ldots, T$, the equilibrium in weakly dominant strategies of the one-shot game obtains at all times.

This amounts to saying that, over a finite horizon, entry occurs in each period, the incumbent being altogether unable to build up a credible entry barrier.

What if the supergame lasts (or is perceived to last) forever? In such a case, it is indeed possible for the incumbent to fight one day and remain a monopolist henceforth. To see this, suppose I and E share the same time preferences, measured by the discount factor $\delta \in [0, 1]$. The discounted profit flow accruing to the incumbent if entry takes place in each period is measured by

$$\Pi_{\mathrm{d}} = \sum_{t=0}^{\infty} \delta^t \pi_{\mathrm{d}} = \frac{\pi_{\mathrm{d}}}{1 - \delta} \tag{5.27}$$

Alternatively, the incumbent may fight at $t = 0$. If this discourages the entrant, by persuading the latter that any entry will trigger an aggressive response (say, a price war),[11] the resulting discounted profit flow for the incumbent is

$$\Pi_{\mathrm{w}} = \pi_{\mathrm{w}} + \sum_{t=1}^{\infty} \delta^t \pi_{\mathrm{M}} = \pi_{\mathrm{w}} + \frac{\delta \pi_{\mathrm{M}}}{1 - \delta} \tag{5.28}$$

Aggressive behaviour is rational if $\Pi_w \geq \Pi_d$, that is, if the following condition is satisfied:

$$\delta \geq \frac{\pi_d - \pi_w}{\pi_M - \pi_w} \equiv \delta^w \in (0, 1) \tag{5.29}$$

On the basis of (5.29), one may draw the conclusive claim contained in the following remark.

REMARK 5.3 If the entry game repeats infinitely many times, the incumbent will fight entry in the first period – retaining thus monopoly power over $t \in [1, \infty)$ – for all $\delta \in [\delta^w, 1]$.

The structure of this supergame replicates that of Friedman's theorem to reach a conclusion that has a completely different flavour as compared to collusive stories. Here, the shadow of the future reveals to the incumbent whether it is convenient to give up a slice of current profits to preserve future ones by adopting an aggressive behaviour on the very first day. Because of complete information, the entrant is able to (i) deduce whether the incumbent has the incentive to adopt this line, and (ii) correctly guess whether to enter or not at the (unique) subgame perfect equilibrium.

Further reading

The literature on repeated games is so large that any attempt to list all the relevant contributions is doomed to failure. An overview of the fundamental steps of this branch of game theory is given in Fudenberg and Tirole (1991). An elegant and reader-friendly introduction to optimal punishments can be found in Gibbons (1992), while exhaustive surveys of applications to oligopoly supergames are given in Martin (1993, 2002). A very interesting and somewhat informal read is Axelrod (1984). For a discussion of antitrust rules in the USA, the EU and Japan, see Bork (1966), Martin (1998) and Iyori and Uesugi (1983).

6   Understanding politics

A politician needs the ability to foretell
what is going to happen tomorrow,
next week, next month, and next year.
And to have the ability afterwards
to explain why it didn't happen.

Sir Winston Churchill

It's not up to me to say whether Sir Winston was right or not, but this quote provides a some-what spirited *introitus* to a chapter dealing with a game-theoretic approach to politics. Not unlike firms, political parties also pursue the attainment of specific goals (winning elections, maximizing consensus, remaining in government for as long as possible) and design their strategies accordingly.

Parties offer political platforms to voters very much like firms offer products to consumers, so that consensus and demand are two closely related phenomena governed by either voters' or consumers' preferences, respectively. Definitely, the whole matter boils down to shares, in both cases. In order to 'sell', parties have to invest money in electoral campaigns that closely resemble firms' advertising campaigns to modify consumers' preferences. The only essential difference is that economic markets operate through a price mechanism that does not appear in politics.

After a synthetic exposition of the astonishing paradoxes that inherently affect voting procedures, I will drive you through an introductory analysis of the strategic interplay tak-ing place in a stylized two-party system, with parties (or their candidates) striving to win elections and gain the office at stake. These games closely replicate the basic features of analogous ones where firms invest in advertising, and also share with the latter the key fea-ture of being socially inefficient, as strategic incentives lead players (be they firms or political parties) to overinvest as compared to what would be both socially and privately efficient.

Relatedly, any administration would definitely prefer to remain in office, and to this end might decide to make strategic use of public announcements in the lead-up to new elections, in order to stay in power. Indeed, there are announcements that we should not take at face value.

6.1   Voting paradoxes

I will set out by telling you a well-known story about a historical event that most of you are likely to be familiar with. In 1976, Jimmy Carter won the US presidential elections against Gerald Ford. The latter had previously obtained the nomination by the Republican party, defeating Ronald Reagan. According to the polls, Reagan would have made it to the

White House (in fact, he did, but later, in 1980). Yet, the Republicans lost the 1976 elections because they preferred Ford to Reagan.

This specific event is an example of a *cyclical* problem affecting majoritarian systems where candidates are matched pairwise, and the final outcome only depends upon the matching order. This phenomenon is known as the *paradox of voting*, or the Condorcet paradox, named after Monsieur le Marquis de Condorcet, who formalized it in 1785. It is not a game, but still it is a pillar of social choice theory, and therefore it must have a place in any book dealing with the formal analysis of political behaviour.

To appreciate what this paradox is about, one first has to understand the definitions of transitivity and transitive relation in mathematics. As an example, take the integers, $1, 2, 3, \ldots, N$. In the realm (or, more rigorously, in the domain) of integers, the relations 'is greater than' and 'is at least as great as' (represented synthetically by the symbols $>$ and $\geq$, respectively) are transitive, as we can correctly state that, taking any three integers A, B and C:

$$\text{if } A > B \text{ and } B > C, \text{ then } A > C \tag{6.1}$$

and, similarly, if $A \geq B$ and $B \geq C$, then $A \geq C$. For instance, the triple ($A = 3$, $B = 2$, $C = 1$) satisfies (6.1). There are of course many relations that are not transitive. One is 'to be the son of': if John is the son of Robert, and Robert is the son of George, then John is not the son of George.

Now let's go back to Condorcet and consider the relation 'to be preferred to', summarized by the symbol $\succ$. I invite you to examine an electoral problem with three voters, a, b and c, and three candidates, α, β and γ. The voters' preferences are as follows:

- a: $\alpha \succ \beta \succ \gamma$
- b: $\beta \succ \gamma \succ \alpha$
- c: $\gamma \succ \alpha \succ \beta$

The above list of preferences is indeed cyclical, and majority voting yields no clear-cut solution. Indeed, should any of the three candidates be selected as the winner, there would arise strong (and sound) objections: if candidate α were chosen, two voters (b and c) would yell that they prefer candidate γ. The very same argument applies to the choice of the other two candidates, and cyclicity results in a three-way tie across candidates, with majority voting leading to a stalemate.

Condorcet proposed a way out of the paradox – known as the Condorcet method – in order for majority voting to yield a winner. In summary, the method consists in pitting every candidate against every other candidate in a series of imaginary one-on-one contests, with pairwise counting. I will not delve into the details of the different procedures that can be adopted to solve the paradox, but this way out is precisely what is being adopted in the US presidential elections and elsewhere – yielding a winner, but not necessarily selecting as such the candidate who would have the largest consensus across the population of voters.

A direct, although much later, follow-up to the Condorcet paradox is Arrow's *impossibility theorem*, or Arrow's paradox (Arrow, 1951). Arrow's theorem proves that, when voters face three or more options, there exists no voting system capable of converting individually ranked preferences into a community-wide ranking while also meeting a certain set of criteria. An equivalent formulation of the theorem is that, in the absence of restrictions on either individual preferences or neutrality of the constitution as to feasible alternatives, there exists no social choice satisfying a set of seemingly plausible requirements. The result generalizes the Condorcet voting paradox.

To make things as simple as possible, one has to think that the subject of Arrow's theorem is the problem of aggregating in an acceptable and satisfactory way the preferences of a number of individuals. This aggregation procedure is supposed to produce a system of what this literature calls social preferences, through some generally accepted mechanism that is labelled as a social welfare function, which has to meet a set of sensible conditions or rules. The complete list of such 'plausible requirements' includes the following.[1]

- *Universality* (or, unrestricted domain): given any set of individual preferences, the common social welfare function should yield a unique and complete ranking of societal choices, in such a way that (i) the ranking of preferences for society is complete, and (ii) the same social ranking of alternatives emerges each time voters' preferences are presented the same way.
- *Independence of irrelevant alternatives*: the social welfare function should rank alternatives α and β in the same order, irrespective of whether one or more additional alternatives (say, $\gamma, \eta, \ldots$) are being evaluated.
- *No dictatorship*: the social welfare function should never reflect the preferences of a single individual (unless in the very special case of unanimity).
- *Monotonicity* (or, positive association of individual and social preferences): this rules out the possibility for an individual to damage any alternative by ranking it higher.
- *Citizen sovereignty* (or, non-imposition): intuitively, this means that any social preference ranking should be achievable starting from some set of individual preferences.

The following theorem is the startling result stated in Arrow's theorem.

THEOREM 6.1 (IMPOSSIBILITY THEOREM) *Given a population of at least two voters facing a set of at least three alternatives, there exists no social welfare function meeting simultaneously the five requirements listed above.*

A useful exercise for you is this: go back to the example used in connection with the illustration of Condorcet's paradox, and try to find out what conditions are violated by it.

So, consider an alternative interpretation of Arrow's impossibility theorem, by considering the historical phase and the context in which Arrow conceived his theorem. At the turn of the 1950s, the core issue was the political and eventually even military confrontation between West and East, between the USA and the USSR. As we already know from Chapter 1, the Cold War prompted an enormous stream of research in several fields, including the social sciences. Many prototypical games are the outcome of that age, and bear given names that are meant to produce a screen of fog between insiders (those involved with the harsh business of learning how to win the Cold War) and outsiders (people like me, working in the ivory tower of the academy). Well, the impossibility theorem can also be examined in this light. One way of reading it is that a dictatorship decides what to do definitely faster (if not necessarily better) than a democratic system, the latter being obstacled by the above 'plausible requirements', including the no-dictatorship condition. An appalling view, isn't it? Of course, this is not reflected in the *vulgata* of Arrow's theorem that circulated in those years. Rather, its interpretation was that a *benevolent dictator* (or a social planner) aiming at the maximization of social welfare could succeed in disentangling socioeconomic problems that would otherwise produce inefficiency and sluggishness if left to individuals with private and reciprocally incompatible goals (in Chapter 4 you have encountered plenty of examples of this nature).

Now that it's clear how problematic it is to choose social welfare functions to aggregate individual votes, let's see how firms behave strategically to gain the largest possible consensus.

6.2 A spatial model of political competition

> The best party is but
> a kind of conspiracy
> against the rest of the nation.
>> Lord Halifax

Do you remember the Hotelling duopoly model I laid out in Chapter 4 to illustrate firms' incentives to differentiate their products and invest in advertising? Take it off the shelf again, strip it of the price-setting stage, and what you get is the spatial model of political competition dating back to Downs (1957, chapter 8).[2]

Examine the following two-party electoral competition game. Players are two parties, labelled *L* and *R* (standing for *Left* and *Right*, respectively). Alternatively, one can think of the linear segment as a measure of the intensity of any given policy, for instance fiscal policy: if so, then point 0 represents a situation where fiscal pressure is absent altogether, while point 1 is the opposite (and purely hypothetical) situation where, say, income taxation is 100 per cent. In this example, I will stick to the idea that the Hotelling–Downs segment represents the spectrum of all possible political colours.

Parties choose their respective platforms $x_R \geq x_L$, with $x_L \in [0, 1/2]$ and $x_R \in [1/2, 1]$.[3] Let the unit interval describe the space of the electors' political preferences, ranging from the extreme left (0) to the extreme right (1). Voters are uniformly distributed over the interval, with density d in each point, so that the total number of votes is $d \cdot 1 = d$. For the sake of simplicity, suppose everyone indeed votes (either for party *L* or for party *R*), so that the total amount of votes is equal to d.[4] The generic voter, located in point $m \in [0, 1]$, votes for the party (or the candidates) whose electoral platform x_i maximizes the following utility function:

$$U = s - t(m - x_i)^2 \qquad (6.2)$$

where $s > 0$ is the gross value that any individual voter associates with the fact itself of voting, while parameter $t > 0$ measures the disutility of voting for a party whose platform differs form the voter's ideal one.

The voter who is indifferent between candidate *L* and candidate *R* is identified by the equation:

$$s - t(\widetilde{m} - x_L)^2 = s - t(\widetilde{m} - x_R)^2 \qquad (6.3)$$

which can be solved to identify the position of such a voter:

$$\widetilde{m} = \frac{x_L + x_R}{2} \qquad (6.4)$$

All this immediately involves the following result.

REMARK 6.1 The votes of the electors located between x_L and x_R split evenly between the two parties or candidates.

It is worth stressing that, in the original Hotelling model with price competition in the market, this result obtains only in the very special case where firms' prices are identical. Yet, what is really interesting of Remark 6.1 is what it doesn't explicitly state, namely, that each party has its own private backyard (or hunting ground), measured by x_L for the Left and $1 - x_R$ for the Right, respectively. In each of these two ranges, the party located to the boundary of it is an unchallenged master, but the ultimate outcome of the elections is determined by the population of voters whose consensus is, so to speak, *contestable*.

A distinctive feature of the model is therefore that each candidate cares only about the outcome of the election – the distribution of votes across the candidates – and, unlike the voters, not about the specific position of the winning candidate. In a sense, candidates and parties do not internalize the external effect their behaviour produces for the community as a whole. What matters is solely winning the elections. In the remainder, I will come back to this crucial aspect.

Now we can proceed to determine the partition of votes, v_L and v_R, for a generic pair of political platforms. All voters to the left of $\tilde{m}$ will vote for candidate 1, while the remainder of them, from $\tilde{m}$ to 1, will patronize candidate 2. Therefore,

$$ v_L = \frac{d(x_L + x_R)}{2}, \quad v_R = d\left(1 - \frac{x_L + x_R}{2}\right) \tag{6.5} $$

From (6.5), it appears that v_L increases with x_L, while v_R decreases with x_R. This amounts to saying that the Left has an incentive to relocate rightwards, while the Right has an incentive to relocate leftwards. By doing so, each party hopes to (i) maximize the size of its own private backyard and, at the same time, (ii) subtract as many votes as possible from the rival in the contestable sub-interval between x_L and x_R. Also note that any symmetric pair of platforms, with $x_L = 1 - x_R$, is such that $v_L = v_R = d/2$. Therefore, there seems to be room here for the median voter at $1/2$ to be playing a pivotal role in determining the winner.

Now take a look at Figure 6.1, which captures the essential features of this game. In particular, the two arrows illustrate the dynamics of both parties along the unit segment. Given the dynamics affecting parties' incentives as to their positions along the segment, their respective initial locations are in fact altogether immaterial.[5]

The relocation tendency affecting both parties alike (although with opposite sign) ultimately leads both of them to design a political platform that exactly fits the preferences of

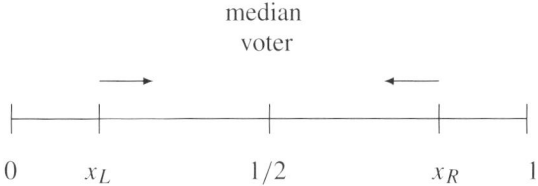

Figure 6.1 The Hotelling–Downs segment

the median voter (again, the 'middleman') located at $1/2$, as summarized in the following remark.

REMARK 6.2   (MEDIAN VOTER THEOREM) At the Nash equilibrium of the Downs game, both parties' electoral platforms coincide with the median voter's.

This simple exercise explains the reason why, in real-world cases, more often than not, platforms are so similar to each other when it comes to relevant issues such as economic policy or foreign affairs. The intuitive interpretation of this outcome can be spelled out on the basis of the demand effect we have met in Chapter 4 while examining the Hotelling model: since here there are no prices, what matters is only to capture the attention of the middleman, and the gravity attraction exerted by the latter univocally determines the behaviour of parties. However, note that, as soon as $x_L = x_R = 1/2$, all voters are in fact totally indifferent as to which party to patronize precisely because Left and Right have become identical in all respects. This entails two related consequences:

- if $x_L = x_R = 1/2$, the median voter is not necessarily pivotal, while (or, because)
- any voter is potentially pivotal.

Shall we take the result of the median voter theorem for granted, i.e., shall we buy it at its face value, or is there more to it? This delicate issue is the subject of the next section.

6.3   The robustness of the median voter theorem

As I have mentioned above, the linear segment may represent the set of all admissible policy stances a party or candidate may take, concerning the pension system, income taxation, etc. If this is the interpretation of the Hotelling–Downs model, then the first possible extension consists in allowing for more dimensions to jointly define a candidate's political platform. It can be shown that the basic result is robust to such a generalization (see Davis *et al.*, 1972; Hinich *et al.*, 1972, 1973; Slutsky, 1975; Austen-Smith, 1983; Calvert, 1985; Ansolabehere and Snyder, 2000), although in this case a pure-strategy equilibrium may not exist – the mixed-strategy equilibrium has been investigated by McKelvey and Ordeshook (1976) and Kramer (1978).

If voters (or their preferences) are not uniformly (or even symmetrically) distributed along the segment, then there is no guarantee that the median voter theorem may hold. The last 50 years of Italian politics constitute an evident illustration of the consequences of polarizing political preferences: while during the so-called *First Republic* (roughly, over the period 1950–1990) the distribution of political tastes was clearly single-peaked and symmetric around the midpoint, the collapse of the Christian Democrats gave way to a polarized set-up lasting to the present day.

The possibility that the median voter may lose his/her role as an attraction point in situations in which it is not the case that every citizen votes with absolute certainty for his/her favourite candidate was first pointed out by Hinich (1977), where the probability that any citizen votes for any given candidate is taken as a primitive, i.e., it is exogenously given (following, in this respect, Hinich *et al.*, 1972). This contribution opened up an alternative approach, known as the *probabilistic voting framework* (see Coughlin, 1990, 1992; Patty, 2005; Schofield, 2006; *inter alia*).

Another, quite sensible, extension of the basic model, which may represent a serious challenge to the appeal exerted by the median voter, is that where each candidate does have an *ideological stance* (an ideal position along the segment) that he/she may modify or abandon only at some cost. This view dates back to Wittman (1977, 1983, 1990) and has been investigated extensively (see Calvert, 1985; Roemer, 1994). The simple game we are about to examine is loosely based on Lambertini (2007).

6.3.1 An electoral game with 'platform stickiness'

The aim of this digression is to extend the traditional spatial approach *à la* Hotelling–Downs in order to investigate the bearings of parties' traditional platforms, as inherited from history, on their strategies and ultimately on the outcome of electoral competition. When I refer to the presence of traditional platforms, I mean the following. Any given party may take from its past history some essential features conditioning its views on relevant policy issues, such as monetary and fiscal policy, welfare, foreign policy, etc., so much so that they end up shaping to a large extent the choice of such a party's platform during the elections. This aspect, close in spirit but not equivalent to Wittman's approach (and its follow-ups), may in fact play a relevant role, as can be easily ascertained on the basis of casual observation.

For instance, a phenomenon of this type clearly emerges if one takes a casual look at the political elections held in Italy in April 2006, with particular regard to the behaviour of the centre-left during the last two weeks of the electoral campaign, when Mr Romano Prodi's public speeches and declarations were increasingly stressing the need to increase fiscal pressure (or introduce new taxes) on large patrimonies and high incomes so as to make the income distribution in Italy less unfair. Exactly how large such patrimonies were supposed to be remained a vague concept until the new government issued the new fiscal law. Yet, these declarations of intent produced the effect (predictable but clearly – and quite strangely indeed – unforeseen by the centre-left itself) of decreasing their margin of consensus over the centre-right coalition to such an extent that the outcome of the elections was pretty tight, being determined by a few thousand votes only. It is a widely accepted interpretation that such declarations were dictated not by a risk-loving attitude but rather from the will (or need) to satisfy some intrinsic (or tradition-driven) requirements of some components of the coalition led by Mr Prodi.

Having motivated the exercise, let's turn to the formal side of it. The model closely reflects the Hotelling–Downs set-up outlined above, with a linear segment [0, 1] along which voters are uniformly distributed with density d, and abstentions are assumed away. However, in the present case there are two novel features that play a role in shaping parties' behaviour and the outcome of elections: historical traditions, and the related costs of moving away from them.

For the sake of simplicity, we may suppose that parties L and R have their respective historically determined locations (or ideological stances) at $1/4$ and $3/4$, respectively, along the unit interval. If each party sticks to its respective historical heritage, then voters split evenly across parties (or candidates), with $v_L = v_R = d/2$, as in the original model, except that here parties are not meeting the tastes of the median voter.

Now, the point is that each party has the possibility of abandoning (or, to some extent, betraying) its historically inherited stance to move rightwards or leftwards to reach $1/2$, responding thus to the gravitational attraction exerted by the median voter. To do so, a party must bear a positive cost C that quantifies the psychological unwillingness to abandon consolidated traditions. Accordingly, the utility functions of the two parties or candidates can be written as follows:

$$u_L = \begin{cases} v_L(1/2, x_R) - C & \text{if } x_L = 1/2 \\ v_L(1/4, x_R) & \text{if } x_L = 1/4 \end{cases} \qquad (6.6)$$

and

$$u_R = \begin{cases} v_R(x_L, 1/2) - C & \text{if } x_R = 1/2 \\ v_L(x_L, 3/4) & \text{if } x_R = 3/4 \end{cases} \tag{6.7}$$

where, for the moment, the rival's location is generic. In order to compute the exact volumes of votes, one has to recall that any voter located in the interval $[0, x_L]$ votes for the Left, while any voter in $[x_R, 1]$ votes for the Right, while the subpopulation of voters in (x_L, x_R) splits evenly. The list of all possible cases is as follows:

- $x_L = 1/4$ and $x_R = 3/4$, whereby $u_L = u_R = v_L = v_R = d/2$;
- $x_L = x_R = 1/2$, whereby $v_L = v_R = d/2$ and $u_L = u_R = d/2 - C$;
- $x_L = 1/2$ and $x_R = 3/4$, whereby $v_L = 5d/8$, $v_R = 3d/8$ and $u_L = 5d/8 - C$, $u_R = 3d/8$;
- $x_L = 1/4$ and $x_R = 1/2$, whereby $v_L = 3d/8$, $v_R = 5d/8$ and $u_L = 3d/8$, $u_R = 5d/8 - C$.

Of course, given that $5d/8 > 3d/8$, unilateral relocation by either party when the rival is sticking to its tradition pays off in terms of the pure electoral outcome, but the costs involved in relocation might well induce each party to remain at the location inherited from the past. The essence of strategic interaction is captured by Matrix 6.1.

Since C plays a key role, we have to see what happens for any admissible level of this cost parameter. The key inequalities in determining the equilibrium outcome(s) of the game are

$$\frac{d}{2} \gtrless \frac{5d}{8} - C, \quad \text{that is, } C \gtrless \frac{d}{8} \tag{6.8}$$

and

$$\frac{3d}{8} \gtrless \frac{d}{2} - C, \quad \text{that is, } C \gtrless \frac{d}{8} \tag{6.9}$$

so that the critical threshold of C is $d/8$ in both cases. Hence, for all $C \in (0, d/8)$, strategy $1/2$ dominates, and the unique Nash equilibrium is $(1/2, 1/2)$, with both parties relocating in correspondence to the median voter. Otherwise, for all $C > d/8$, the unique equilibrium is such that both parties stick to their respective traditional platforms.

Additionally, since obviously $d/2 > d/2 - C$ for all values of C, then, whenever the pair $(1/2, 1/2)$ is the equilibrium, the game is a prisoners' dilemma (i.e., the equilibrium is Pareto-inefficient as the parties bear the costs of relocating towards the median voter, to no avail). Conversely, if the equilibrium is $(1/4, 3/4)$, this game is *never* a prisoners' dilemma, in that the parties avoid wasting resources and the equilibrium outcome is indeed Pareto-optimal.

Matrix 6.1 The 'platform stickiness' electoral game

		R	
		3/4	1/2
L	1/4	$d/2; d/2$	$3d/8; 5d/8 - C$
	1/2	$5d/8 - C; 3d/8$	$d/2 - C; d/2 - C$

Matrix 6.2 The game with asymmetric
stickiness

		R	
		3/4	1/2
L	1/4	$d/2; d/2$	$3d/8; 5d/8$
	1/2	$5d/8 - C; 3d/8$	$d/2 - C; d/2$

This fully symmetric model yields alternative but still fully symmetric equilibria. In order to observe asymmetric equilibria one has to modify the set-up so as to allow for the possibility that one of the two parties or candidates does not bear any relocation costs – that is, tradition is not a burden to worry about, either for the Left or for the Right.[6] Let's say this holds for the Right. In such a case, the game modifies as in Matrix 6.2.

Now strategy 1/2 is always dominant for the Right, while the Left has to evaluate the same inequalities as above. Hence, the game yields a unique equilibrium in strictly dominant strategies, the equilibrium outcome being either (1/2, 1/2) for all $C \in (0, d/8)$ or (1/4, 1/2) for all $C > d/8$.

To sum up, I state the following remark.

REMARK 6.3 If historical heritage matters exactly the same for both parties, equilibrium platforms will both coincide with the median voter's tastes provided that relocation is not too costly. If there arise asymmetries in relocation costs, then the party whose perceived costs are lower will always satisfy the median voter's preferences, while the other may not, thus giving rise to an asymmetric platform pair at equilibrium.

6.4 Electoral campaigns

> Politics is not a game.
> It is an earnest business.
>   Sir Winston Churchill

In Chapter 4, we have seen how strategic interaction may lead oligopolistic firms to carry out excess investment in advertising. The very same phenomenon can be observed as well in electoral competitions where parties invest in long and costly campaigns to conquer the citizens' votes.

The simplest way of grasping the intuition as to why the huge amount of resources (both time and money) invested in electoral campaigning turns out to be both privately and socially wasteful is to build up a highly stylized 2×2 game as in Matrix 6.3.

Parties *L* and *R* choose between two pure strategies, either investing or not a given amount of money *k* to finance the electoral campaign. As a benchmark (which, by the way, has often proven to be not unrealistic), let's say that, in the absence of any investment by parties, the

Matrix 6.3 Wasteful efforts in electoral campaign

		R	
		0	k
L	0	50%; 50%	25%; 75%
	k	75%; 25%	50%; 50%

shares are symmetric, 50/50; and, in the asymmetric event, in which one party undertakes the costly campaign while the other does not, this turns out to be decisive in terms of the final outcome: the candidate of the party that invests outperforms the rival and ultimately gets into office. The most problematic aspect of this story is that the remaining symmetric outcome where both parties invest yields exactly the same shares as the one where no spending occurs at all (on purpose, I also chose the shares corresponding to this outcome so as to remind you about events you may have observed in recent times). The underlying hypothesis on which this game is based is that parties raise external funds, say, from DrinkThis and TasteThat, respectively, so that neither party really cares about expenditures to finance the campaign, and the payoffs are solely specified in terms of consensus. Hence, Matrix 6.3 portrays once again a prisoners' dilemma, whose unique equilibrium in dominant strategies (k, k) indeed entails wasting $2k$ without modifying the *ex ante* distribution of consensus across candidates. Note that, leaving the shares of votes totally unmodified, this investment is inefficient in all respects, i.e., both privately (for parties) and socially (i.e., from a collective standpoint).

6.5   How about being re-elected?

> The government's view of the economy
> could be summed up in a few short phrases:
> If it moves, tax it.
> If it keeps moving, regulate it.
> And if it stops moving, subsidize it.
>
> Ronald Reagan

So far, I have reviewed games where candidates want to be elected. Another relevant matter is to understand what they may do in order to stay in office when elections approach. This observation prompts the analysis of an intriguing and potentially very dangerous endogenous link between the design of any given policy (not necessarily of economic nature only) and political consensus. A government can purposely design economic policy, or make announcements about future economic measures, in order to be re-elected. The key question is therefore whether these policies or announcements can be deemed credible or not. This is a hot topic in the literature on macroeconomic policy and has received attention ever since the 1980s,[7] being closely related to the debate on rules versus discretion that we have run across in Chapter 4 (see, in particular, Matrix 4.11).

The game we are about to look at is one where the two players involved are the party in government (G) and the citizens/voters (V), who may be relabelled as 'the economic system'. To keep things simple, let's say that the policy menu is one-dimensional, the task of the government being limited to determine the intensity of fiscal pressure on the aggregate income generated by the economic system. The gross domestic product (GDP) is Y. The government has two options: either to keep taxation at a given level T, or to eliminate it altogether (strategy 0). Now suppose elections are approaching, and the premier publicly announces that, if he/she gets re-elected, income will become tax-free. Voters may either believe this announcement and confirm the current administration in office (strategy b), or not (strategy nb). In the latter case, the rival party will win the next elections and design fiscal policy according to its likings or needs. If voters have no specific information about such likings or needs, they may attach equal probabilities to the two mutually exclusive events, 0 and T.[8]

Matrix 6.4 A game of announcements
and credibility

		b	nb
G	0	$S; Y$	$0; \frac{1}{2}Y + \frac{1}{2}(Y - T)$
	T	$S + T; Y - T$	$0; \frac{1}{2}Y + \frac{1}{2}(Y - T)$

Voters' utility is proportional to consumption C, which in turn is proportional to disposable (i.e., after tax) income:

$$C = c(Y - T) \qquad (6.10)$$

where parameter c is positive. On the other hand, the objective of the government is a combination of two elements: to stay in office, S, and to raise money through taxes, $T \in (0, Y)$ (say, for financing public expenditure). Accordingly, this game can be represented through Matrix 6.4.

The government is the row player while voters are the column player. Observe that the two outcomes appearing along the second column coincide, as the voters are not buying the announcement and therefore the other party wins the elections. Also note that voters have no dominant strategy, since

$$Y > \frac{1}{2}Y + \frac{1}{2}(Y - T) > Y - T \qquad (6.11)$$

while the government has in T a weakly dominant strategy. So, in the absence of any credible commitment device to tie the hands of the government, we may expect a reconfirmed administration to use the fiscal instrument at will no matter what was the message conveyed by *ex ante* announcements. This latter piece of information is the key to the solution of the game. After deleting the first row of the matrix, nb becomes dominant for voters, so that we expect to observe a change in the administration, (T, nb) emerging as the equilibrium in (weakly) dominant strategies, attainable by iteration.

There is more to it: the most sensible way of figuring out how a game like this works is to think of it in terms of sequential play, with the voters moving first. After all, they have to decide whether to confirm in office or not the current administration, which later, if still in office, will have to decide what to do with the tax bill. Figure 6.2 illustrates the Kuhn tree of the game, with voters at the root.[9]

Proceeding by backward induction, we may see that G's best reply to b is to tax the economy. Voters are well aware of this consequence because of complete information, and consequently will opt for nb. The sequential play interpretation of the game thus yields (T, nb) as the subgame perfect equilibrium under perfect information.

Essentially, what the extensive form adds to the picture is the interpretation of the fundamental story told by the game: once re-elected, the current administration will focus on raising money through taxes, forgetting about any previous announcement that was solely designed to achieve re-election. Anticipating this unpleasant consequence, and assessing the non-credible announcement for what it really is, voters prefer to take a fair chance with the opposition. To sum up, I may draw the following conclusion (quite alarming, alas).

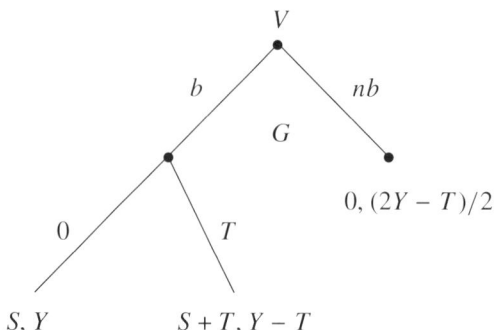

Figure 6.2 Announcements and credibility under sequential moves

REMARK 6.4 Voters should not buy at face value any announcements not sustained by credible commitments, as they are doomed to be systematically reneged upon.

Further reading

The issues treated in this chapter can be further investigated by having a close look at Brams (1975, 1978), Ordeshook (1986, 1992), Morrow (1992) and Osborne (1995). A more detailed (and a little bit more formal) treatment of the voting paradox can be found in Montet and Serra (2003). A very nice read on the relationship between economics and politics is Dunleavy (1991). There are further – virtually uncountable – relevant contributions adopting a spatial approach to investigate other aspects of electoral competition: see, e.g., Bartholdi *et al.* (1991), Weber (1997), Adams (2000), McKelvey and Patty (2006) and Huck *et al.* (2006). For an analysis of probabilistic voting, see Adams (1999).

7 Wargames

We may anticipate a state of affairs in which two Great Powers will
each be in a position to put an end to the civilization and life of the other;
though not without risking its own. We may be likened to two scorpions
in a bottle, each capable of killing the other; but only at the risk of his own life.

Robert J. Oppenheimer

With this chapter, we enter the realm of international relations. Here, in particular, we will deal with the class of problems that so strongly motivated research activity in game theory in its early days.

Given the nature of the topics at stake and the size of the extant literature, I will be unable to cover the existing material exhaustively, but I will give you what I hope will prove to be a stimulating perspective on a set of crucial and far-ranging issues that remain the focus of the political scientists' attention.

As an appetizer, we'll start with two games describing important episodes of World War II. These games are representative of the effort made by game theorists at the outset of the Cold War to 'sell' game theory to decision-makers. To do so, applied research in this field had to prove itself able to reconstruct historical facts with a high degree of precision.

In the remainder of the chapter, I will review several games sharing the same *leitmotiv*, namely, escalation. These are games that have been elaborated – if not necessarily in the form you will see here – during the crises they were meant to interpret. Accordingly, the second part of the chapter brings to the fore the predictive or normative power of the theory.

The first of these escalation games is not really a game but an auction, with a highly stylized model being chosen with the explicit purpose of illustrating that there are situations in which one would better never ever get involved. The following two games portray the Cuban missile crisis and the European one, both critical turning points of the Cold War. The last set-up models an aspect that is still of interest nowadays: the possibility for a nation to endow itself with a shield against nuclear strikes. The specific feature of this latter game consists in shedding some light on the question whether supposedly defensive instruments are indeed such or, instead, they have a quite different nature.

7.1 The battle of the Bismarck Sea

This game dates back to Haywood (1954), and offers a reconstruction of the military confrontation between US and Australian forces on one side and Japanese ones on the other for possession of the Japanese base of Lae, located at the eastern end of New Guinea, during World War II in the Pacific theatre (February–March 1943).

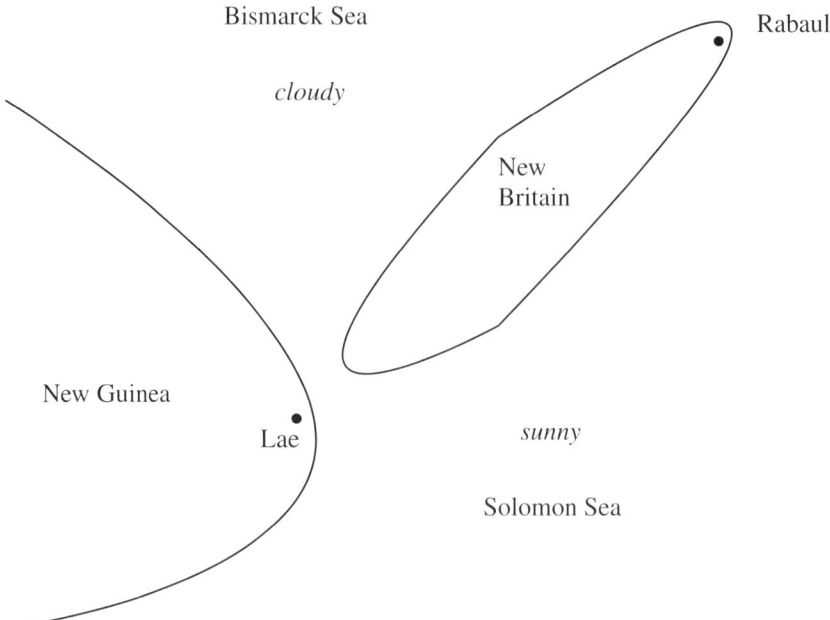

Figure 7.1 Schematic map of the battle of the Bismarck Sea

Since Lae was under siege and about to fall, the Japanese headquarters decided to try to reinforce it by sending a naval convoy with troops and supplies, sailing from the largest base remaining in Japanese hands in the Pacific, Rabaul, at the eastern end of New Britain.

The Japanese expected a reaction in the form of air strikes, and had to figure out how to escape it. The convoy could sail along either the north route or the south route, both requiring approximately three days to reach Lae from Rabaul. The only difference between them was the weather, which was sunny along the south route and conversely cloudy along the north one. Hence, bad atmospheric conditions could help the Japanese ships to sneak through and relieve the sieged garrison. On the opposite side, the Allied expected the Japanese to send resupplies to Lae (based on intelligence reports) and had to figure out the Japanese strategy and try to intercept the convoy, based upon the very same information about the weather forecast. A sketch of the situation is illustrated in Figure 7.1.

The strategies available to Allied bombers and the Japanese convoy are *north* (N) and *south* (S), and the strategic form of the game can be laid out on the basis of the expected duration of the bombing raids. The players' *ex ante* conjectures can be summarized as follows:

- If both players choose S, that would result in complete annihilation for the Japanese, as the Allied aircraft would immediately spot the convoy and bomb it for three days, thanks to good weather.
- If both choose N, the bombing raids would last for two days only because one would be eaten up by bad weather hiding Japanese ships.
- If the Allied Command should initially send its bombers southwards while the Japanese convoy is actually sailing along the north route, only a one-day window would be left

Matrix 7.1 The battle of the
Bismarck Sea

		JC	
		N	S
AB	N	2	2
	S	1	3

open for raids, one day being wasted by the bombers to redirect north, another day being
again wasted due to bad visibility.

- Finally, should the bombers initially go north while the convoy is already following the
  south route, two days would be available to Allied bombers to strike the convoy, as they
  only waste one day looking for the ships north of New Britain.

The strategic form of the game is thus as in Matrix 7.1. Players are the Allied bombers (*AB*)
and the Japanese convoy (*JC*), and information is imperfect, complete and symmetric.

Since payoffs are measured by the expected number of bombing days, the Allied forces
want to maximize them, while the Japanese want to minimize them. Also, note that the
present model is a constant-sum game (i.e., strictly competitive) and therefore one may indif-
ferently solve it using either the Nash equilibrium or the maximin (or minimax) criterion *à la*
von Neumann. Additionally, a quick look at the matrix reveals that *N* is a weakly dominant
strategy for the Japanese convoy. Hence, after deleting the left column, *N* emerges as a dominant strategy for the Allies. As a result, the game yields a unique Nash equilibrium in weakly
dominant strategies, (*N*, *N*), attainable by iteration. This outcome coincides with the facts,
the Japanese convoy being disastrously bombed by Allied aircraft. And, as a consequence of
the lack of resupplies and reinforcements, Lae fell.[1]

This example illustrates the general mood characterizing the work of game theorists as
well as other scientists operating with the RAND Corporation and ONR immediately after
the end of World War II. The possibility of reconstructing well-known historical events as
games or, more generally, formal models, was the prerequisite for a more ambitious project,
where the same tools were to be used so as to suggest to the executive power the best strategy
in virtually any possible critical scenario. That is to say, proving the descriptive power of this
(or any other) theory was the necessary condition to consolidate the view that the same theory
also had a reliable normative capability.

7.2 Overlord

> Take calculated risks.
> That is quite different from being rash.
> > George S. Patton

As many of you may know, General Patton was not directly involved in the amphibious
landings of Overlord, but he indeed played a key deception role to induce the Germans to
believe that the objective of the invasion force that was due to cross the English Channel
was not Normandy but rather the Pas de Calais. Correctly anticipating the location of Allied
landing zones was the biggest question for the Germans to answer (besides of course the
situation along the Eastern Front). Of course, the most obvious option was the Pas de Calais,
but this was even too clear to both Allies and Germans. Therefore, the Allies could indeed
prefer the longer route leading to the beaches of Normandy. The game we are about to see,

Matrix 7.2 The Overlord game:
Normandy or Calais?

		Germans	
		C	*N*
Allies	*C*	20; 80	100; 0
	N	80; 0	60; 20

building upon Dixit and Nalebuff (1991, chapter 7), describes the strategic interplay between the Allies, choosing where to direct their strike, and the Germans, choosing where to locate the scarce resources to oppose the landing.[2]

Define as {*C, N*}, standing for *Calais* and *Normandy*, respectively, the set of admissible pure strategies for both players, Allies and Germans. The *a priori* probabilities of success for the Allies are the following:

- 80 per cent if both players choose *N*;
- 20 per cent if both players choose *C*;
- 100 per cent if the Allies choose *N* while the Germans choose *C*;
- 100 per cent if the Allies choose *C* while the Germans choose *N*.

This means that, if the Germans concentrate their forces in the 'wrong' place, the invasion succeeds with certainty; otherwise, if the Germans choose the 'right' locations, the Allies' chances are higher in Normandy than at the Pas de Calais.

To build up the associated payoffs appropriately, suppose that the Allies attach a higher value to landing at the Pas de Calais than in Normandy (because the former is closer to Germany than the latter). So, a successful landing operation at the Pas de Calais is worth 100, while an analogous event in Normandy is worth only 80. Given the prior probabilities of a successful invasion, the Allies' payoffs in the symmetric cases are $0.75 \times 80 = 60$ in (N, N) and $0.2 \times 100 = 20$ in (C, C). The payoffs for the Germans are defined residually, so that in (C, C) the Germans get $100 - 20 = 80$, while in (N, N) they get $80 - 60 = 20$, and so on.

Accordingly, the strategic form of the game is shown in Matrix 7.2. As for the previous game, this is also a constant-sum one, and therefore we may use either the Nash equilibrium or von Neumann's minimax criterion to solve it.

It appears at first sight that the game yields no equilibrium in pure strategies, and consequently one has to resort to mixed strategies. The Allies attach probability p to strategy *C* (and the complementary probability $1 - p$ to strategy *N*), while the Germans attach probability q to strategy *C* (and its complement $1 - q$ to strategy *N*).

From Chapter 3, we know that the Nash equilibrium in mixed strategies obtains by requiring each player to set probabilities so as to make the rival indifferent over the entire set of his/her pure strategies. In the present game, this means that the following condition must hold for the Allies:

$$20q + 100(1 - q) = 80q + 60(1 - q) \tag{7.1}$$

while the following must hold for the Germans:

$$80p = 20(1 - p) \tag{7.2}$$

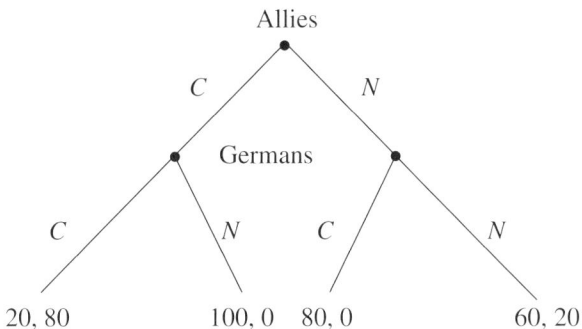

Figure 7.2 The Overlord game in extensive form with perfect information
for the Germans

Solving the above system of equations with respect to q and p we get

$$p^* = \frac{1}{5}, \quad q^* = \frac{2}{5} \tag{7.3}$$

i.e., the pair of optimal probabilities identifying the mixed-strategy Nash equilibrium of
the Overlord game. As you see, this result implies that the level of strategic uncertainty
is higher for the Germans than for the Allies, as the latter *almost certainly* choose N, while
the Germans are not so far away from a 50/50 probability distribution. Relatedly, it is worth
stressing that, although the Germans can reconstruct the above calculations, this fact *per se*
does not help them so much, does it?

Thus far, the game exactly replicates the set-up in Dixit and Nalebuff (1991). To expand
and straighten the above argument, we can think of this game as one unravelling under
sequential play, i.e., perfect information. In particular, let's say that the Allies are moving
first, choosing the beach for the landing: after all, it's up to Eisenhower and Montgomery to
make the first move. This situation is illustrated in Figure 7.2.

This interpretation of the Overlord game relies on the property illustrated at the end of
Chapter 3, whereby a game that appears to have no Nash equilibria in pure strategies under
imperfect information does yield a pure-strategy subgame perfect equilibrium under perfect
information. The Kuhn tree in Figure 7.2 reveals that, if the Allies choose C, the best reply
for the Germans is to choose C. *Mutatis mutandis*, in the other subgame the strategies remain
symmetric, as N is the best reply to N. Therefore, because $60 > 20$, the Allies opt for landing
in Normandy, and the subgame perfect equilibrium in pure strategies is (N, N).

Having reviewed two games offering *ex post* recollections of historical episodes, we are
now acquainted with the descriptive power of games. Hence, we are ready to deal with other
examples designed with a much higher ambition, that of probing the predictive power of
game theory in the middle of a crisis.

7.3 Escalation as an all-pay auction

An arms race is a hall of mirrors

Richard Rhodes,
Dark sun

In Chapter 4, I briefly discussed auction theory in relation to price competition. As you may remember, auctions commonly end up with a winner who pays the first or second price (depending on the structure of the auction), while losers go their way unscathed (if one overlooks the non-negligible fact that they didn't win). What I didn't tell you in Chapter 4 is that there are indeed other auctions, characterized by the property that losing bidders also pay positive amounts. Typically, there exist two alternative auction forms with this feature, the *war of attrition* and the *all-pay auction*. These auction forms share the common feature that all losing bidders pay their bids exactly and differ only in the amounts paid by the winning bidder. In the former, the winning bidder pays the second highest bid, while in the latter the winner pays his own bid. Therefore, they are analogous to standard second-price and first-price auctions, respectively, except that the auctioneer collects all bids at the end.

The archetypal escalation game is the prisoners' dilemma illustrated by Matrix 2.2, which describes the very basics of an arms race. The same issue can also be characterized as an all-pay auction. This is precisely the subject of the famous *dollar auction game* formulated by Shubik (1971).

The structure of the model is the following. Two players, 1 and 2, take part in an all-pay auction where the auctioneer auctions off a single dollar bill. The starting bid is one cent, and the step is one cent as well. The auction ends if no bid is made for a pre-specified amount of time, and ties are ruled out. Observe that, under these rules, the bidding escalation may never cease.

Now suppose player 1 opens up by bidding one cent for the dollar. If 2 doesn't speak, 1 wins, pays the cent and walks away with the dollar, having effectively won 99 cents. What happens if player 2 outbids 1, offering two cents? If 1 doesn't reply, 2 wins, but 1 pays a cent to the auctioneer. So, 1 takes his/her chances and offers three cents. This triggers a further increase to four cents by player 2, and so on. Going on like this, the chain of increasing bids will turn out as a sequence of odd numbers for 1 and even numbers for 2.

This process exhibits two critical stages. The first pops up as soon as the two players' bids are equal to 49 and 50 cents, respectively. Player 2 is winning, but there is an incentive for player 1 to outbid the rival by offering 51 cents for the dollar. The critical nature of this stage of the auction lies in the fact that, at such a moment, the overall sum crosses the break-even point for the auctioneer. The second crucial stage is reached when bids are 99 cents and one dollar, respectively. If player 1 stops bidding, player 2 wins the dollar and breaks even, but player 1 actually loses 99 cents. Since losing one cent is definitely better than losing almost a dollar, player 1 will bring the highest current bid beyond the intrinsic value of the object being auctioned, by offering one dollar and a cent. This, of course, will meet a symmetrical reaction by player 2, and beyond this point both players will surely be losing increasing amounts of money.

In fact, the escalation process may keep going until players reach their reservation prices, which can be thought of as the total amount of their financial resources. In Martin Shubik's own words (Shubik, 1971, p. 111):

> This simple game is a paradigm of escalation. Once the contest has been joined, the odds are that the end will be a disaster to both. When this is played as a parlor game, this usually happens.

This is definitely a type of game people and nations should not play. However, history shows that they did play escalation games, several times.[3] The following two games portray two well-known episodes of the Cold War, where to escalate or not was the question.[4]

7.4 Mutually assured destruction and the Cuban missile crisis

> Khrushchev reminds me of the tiger hunter
> who has picked a place on the wall to hang
> the tiger's skin long before he has caught the tiger.
> This tiger has other ideas.

John F. Kennedy

As we know from Chapter 1, researchers at RAND and ONR were involved in the analysis of the Cold War as well as, to some extent, in the decision-making process that ultimately shaped the behaviour of the White House. The same holds for their peers on the other side of the Iron Curtain. The following games are not the outcome of *ex post* recollections of facts. They were instead designed to interpret critical situations as they deployed.

Escalation was a *leitmotiv* of the confrontation between the USA and the Soviet Union in the second half of the twentieth century, involving not only the armaments race in itself but, parallel to it, the evolution of reciprocal attitudes about the effective use of such arsenals. Here I will lay out a version of a famous game labelled as *MAD* (*mutually assured destruction*), capturing this aspect in the crudest way.[5]

This game portrays the Cuban missile crisis of October 1962, when the Kennedy administration had to face the threat posed by Soviet medium-range missiles with nuclear warheads based in Cuba and almost ready for launching onto most of the US metropolitan territory.

Kennedy chose the route of a limited escalation in the form of a naval blockade, which was further refined into a selective quarantine, contextually presenting Khrushchev and the Kremlin establishment with the perspective of an open nuclear confrontation that would ultimately lead to reciprocal annihilation. After 13 days, at the climax of the crisis, the Soviet Union stepped back, agreeing to withdraw its missiles from Cuba in exchange for an analogous move on the part of the USA, involving the removal of its Jupiter missiles based in Turkey (the latter being comparatively older and already planned to be replaced by much more effective submarine-launched weapons).

The strategic form of the game is given in Matrix 7.3. Let the USA and USSR be the row and column players, respectively. The *status quo ante* upon which the game is based is that the USSR has set up some missiles in Cuba, thus sparking off the crisis. The strategies available to the US administration are either to *accept the status quo* (a) or *escalate* (e) by adopting an aggressive behaviour. The strategies for the USSR are *withdrawal* (w) or escalation (e).

The figures measuring payoffs appearing in the cells of the matrix – although they may look arbitrary – are indeed chosen so as to describe the situation appropriately. First of all, note that $-NW$ (the symmetric payoff accruing to both players in the case of nuclear war) is negative and very large in absolute value. Second, each outcome appearing in the remaining cells can be interpreted as follows.

Matrix 7.3 MAD and the Cuban missile crisis

		USSR	
		w	e
USA	a	$-1; 1$	$-1; 1$
	e	$10; -10$	$-NW; -NW$

- If the USA accepts the deployment of Soviet nuclear weapons in Cuba without any reaction at all, then the distinction between withdrawal and escalation becomes immaterial to the Kremlin, as the USSR has got what it was aiming for, i.e., the creation of a strategic advantage for itself. This is reflected in their respective payoffs along the first row.
- If, instead, the US administration opts for some form of escalation, then the choice between withdrawing or escalating symmetrically matters: in particular, a withdrawal entails a cost to the USSR that outweighs the advantage it could possibly catch along the first row. The alternative, of course, would be the end of the world as we know it (in which case, most likely I wouldn't ever have written this book and you wouldn't be reading it).

In Matrix 7.3 there are two Nash equilibria in pure strategies, (a, e) and (e, w). However, the first relies on the adoption of a non-credible strategy by the USSR, because for the latter strategy e is weakly dominated by strategy w. Since the White House knows it, as the game takes place under complete information concerning strategies and payoffs, by adopting strategy e this may force the Kremlin to withdraw. An alternative but equivalent way of putting it is that, once the USA has chosen e, the USSR has to (i) accept that the first row of the matrix has evaporated and consequently (ii) choose w.

We may sum up these considerations in the following remark.

REMARK 7.1   The outcome (e, w), besides being a Nash equilibrium, is also a (weakly) dominant strategy equilibrium attainable by iterated deletion of dominated strategies, and therefore qualifies as more credible or robust than (a, e).

The extensive form of the game will allow us to appreciate even better the crucial difference between the two Nash equilibria, and the reasons why one should expect the outcome (e, w) to arise as the actual solution of the game. The Kuhn tree of the game is shown in Figure 7.3.

Players move sequentially, with the USA at the root and the USSR operating at the intermediate nodes under perfect information. This is the intuitive way of interpreting sequential interaction, since if the USA accepts the presence of Soviet nuclear weapons in Cuba as a *fait accompli*, then there is indeed no further interplay to wonder about. For this reason, the right subgame (corresponding to the first row of Matrix 7.3) is truncated.

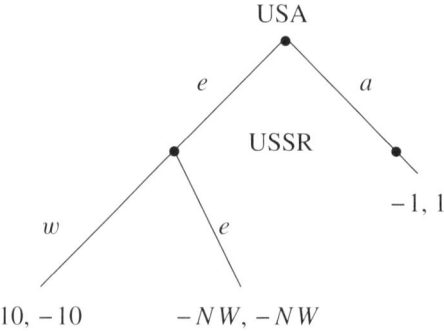

Figure 7.3 The extensive form of the Cuban missile crisis game

This makes it apparent that the Nash equilibrium (a, e) can be reached only if the USA adopts a fearing that the USSR, being put under pressure, goes all the way to the bitter end, engaging in a bilateral escalation process that leads to a nuclear holocaust. However, the Kremlin may use the threat of adopting e as a leverage to solve the crisis in its own favour. If the White House dismisses any reference to this regard as a non-credible announcement, and forces the USSR to proceed along the left subgame by choosing e in order to unmask the Soviet bluff, this selects (e, w) as the subgame perfect equilibrium of the game in extensive form, mirroring what really happened at the end of the 13 days.[6]

This is no walkover, though. The behaviour of the White House during the Cuban missile crisis relies on the idea that the adversary is a rational agent, at least rational enough to avoid mistakes and step back in the face of the ultimate consequences of escalating. The existence of multiple equilibria and the related presence of a mixed-strategy Nash equilibrium reveal how risky a gamble this was for the players (both, not only the US administration). Once probabilities are taken into account, can one rule out the chance of a nuclear confrontation?

To appreciate this aspect of the game, suppose that the Kremlin attaches probability q to strategy w (and its complement $1 - q$ to strategy e). The USA is indifferent between a and e if

$$-q - (1 - q) = 10q - NW(1 - q) \tag{7.4}$$

yielding

$$q^* = \frac{NW - 1}{NW + 10} \tag{7.5}$$

which belongs to the unit interval for all $NW > 1$ (and this has to be true as NW is supposed to be very large). Moreover,

$$q^* - (1 - q^*) = \frac{NW - 12}{NW + 10} > 0 \tag{7.6}$$

This result shows that, indeed, the White House could deem the event of the Kremlin adopting w as more likely than the alternative, *but the latter could not be dismissed a priori with absolute certainty.*

7.5   The Euromissiles crisis

> I do not believe that civilization will be wiped out
> in a war fought with the atomic bomb.
> Perhaps two-thirds of the people
> of the earth will be killed.
>
> Albert Einstein

Back in the 1970s, in the Brezhnev era, the Red Army developed a new type of medium-range nuclear missile, the SS-20, designed to reach Rome, Paris, Bonn or London from its mobile ramps located in or even beyond the Urals (their maximum range being 4400 km). In other words, this missile was a 'theatre' weapon: resorting to the type of rhetorical masterpiece that was commonly in use in those years, it could not be classified as a 'strategic' weapon because it could not reach US metropolitan territory. In view of this feature, the

SS-20 was tailored to trigger and fight a nuclear war confined to Europe. As such, it represented an extremely dangerous instrument, in that it opened a new scenario in which both Superpowers could sacrifice their respective allies without being directly involved in a global thermonuclear war. All this amounts to saying that, while intercontinental (or submarine-launched) ballistic missiles where an instrument of deterrence and, as such, constituted a defensive weapon, the so-called *Euromissiles* (as the SS-20s quickly became known to the public) had an inherently offensive connotation.

The deployment of the SS-20s was viewed by the European members of NATO as an attempt on the part of the USSR to create a rift between them and the USA (the so-called *decoupling strategy*), which would have led to acceptance of the idea of losing Western Europe either to an invasion carried out through the overwhelming conventional forces of the Warsaw Pact or as a consequence of a circumscribed nuclear war fed by Euromissiles on both sides. That was a very delicate moment, as America was coping with the post-Vietnam crisis and many believed that the odds were in favour of the Soviet Union more than ever before.

At first, the US administration seemed keen on a quasi-symmetric reply based on the construction of the neutron bomb, an extremely sophisticated and expensive piece of weaponry lethal for human beings but supposedly not so aggressive for the environment as traditional nuclear weapons. The strategic approach worked out anew specifically for it established that it was to be employed against massive concentrations of enemy armour, if at all possible when the latter were still forming up within the borders of the Warsaw Pact.[7] The Soviet Union had neither the economic nor the technological capabilities needed to build up the equivalent themselves, and felt somewhat uneasy about the prospect of facing a number of neutron bombs in Europe. A lively debate about these weapons arose in Germany and elsewhere in Western Europe, and shortly afterwards President Jimmy Carter stopped the related programme without even negotiating it with Moscow. This unilateral decision initially led the Kremlin to believe that decoupling was indeed attainable.

The game envisaged by Leonid Brezhnev and his associates is represented by Matrix 7.4. Pure strategies *em* and *nem* express the simple dichotomy between having some Euromissiles or not. As to the resulting payoffs, *NWE* stands for *nuclear war in Europe*, while *GNW* stands for *global nuclear war*. Intuitively, it must be *NWE < GNW*, so that the game has two pure-strategy Pareto-rankable Nash equilibria, (*em, em*) and (*nem, nem*), the first being in weakly dominant strategies and Pareto-optimal.[8] On this basis, the Kremlin was trying to convince the White House to share its inclination towards a European conflict that would turn out to leave Western Europe (admittedly, an awfully devastated one) in the hands of the Red Army.

Once again, as in the Cuban missile crisis, the USSR was looking forward to hanging the skin of the tiger in the living room without asking the tiger its own opinion. And, once again, the USSR would have to step back, although in this case it would take much longer than it had in October 1962.

Matrix 7.4 The Euromissiles game as seen from the Kremlin

		USSR	
		em	*nem*
USA	*em*	$-NWE; -NWE$	$-GNW; -GNW$
	nem	$-GNW; -GNW$	$-GNW; -GNW$

Matrix 7.5 The Euromissiles game as seen from the
White House

		USSR	
		w	*ho*
USA	*em*	−NWE; −NWE	−NWE; −NWE
	nem	100; −100	−GNW; −GNW

To see why it happened, we have now to take the viewpoint of the White House. Given that the crisis had been opened up by the Soviet Union (i.e., *there were* SS-20s ready to launch in the Urals), the executive in Washington was looking at a game that did not reflect Matrix 7.4, but rather Matrix 7.5.

While the USA has to decide where to install its own Euromissiles or not (say, across Western Europe), the issue for the USSR is whether to withdraw (*w*) and ultimately dismantle its own SS-20s, or hold out (*ho*), possibly to the bitter end. The payoffs associated with the outcome (*nem, w*) express the idea that a withdrawal would largely diminish the role and influence of the Soviet Union and increase those of the USA at a global level.

Let's suppose the following:

$$100 > -100 > -NWE > -GNW \qquad (7.7)$$

If so, then the game has two Nash equilibria, (*nem, w*) and (*em, ho*), the latter meeting the tastes of the Kremlin. However, strategy *ho* is weakly dominated because of the above chain of inequalities. Hence, the adoption of *ho* is not credible. Conversely, (*nem, w*) lies along the Soviet Union's weakly dominant strategies and is attainable by iteration. Thus, (*nem, w*) is more credible than (*em, ho*) because of the dominance criterion, and coincides with the historical outcome of the crisis, with the USSR agreeing to withdraw and dismantle its problematic SS-20 arsenal.

The intuitive interpretation of this result is that, by rejecting the perspective of nuclear confrontation in Europe, the USA put the USSR in the position of being held responsible for the ignition of a nuclear holocaust, as any strike on a European member of NATO would have necessarily triggered a response based on the only available instrument, namely the intercontinental ballistic missiles (ICBMs) based in North America and their equivalent on board nuclear submarines. The US refusal to get involved at all in the Euromissiles adventure suddenly turned the SS-20s into a costly, useless and risky hardware to be thrown in the garbage bin as soon as possible.

As a compensation, the USA withdrew the Pershing II and first-generation cruise missiles installed in several NATO bases in Europe. This issue was debated, as some (politicians, the press and part of public opinion) considered these weapons as the one-to-one counterpart of the SS-20s. This view is questionable, as the range of such armaments was much shorter than the 4000 km or more attributed to the SS-20s. Moreover, their payload was smaller, as the single warhead of a Pershing varied between 5 and 50 kton,[9] while each SS-20 could carry up to five warheads, each equivalent to 250 kton.

Be that as it may, there are other relevant considerations to keep in mind, and I will briefly outline them, although they stretch well beyond the scope of an introductory text like the present one. Escalation in Europe during the Cold War consistently followed a precise rule, of which both players were well aware. That rule maintained that, in view of the NATO inferior endowments of conventional armaments, the deterrence against a conventional strike by the Warsaw Pact should rely on tactical (i.e., short-range) nuclear weapons. That is to

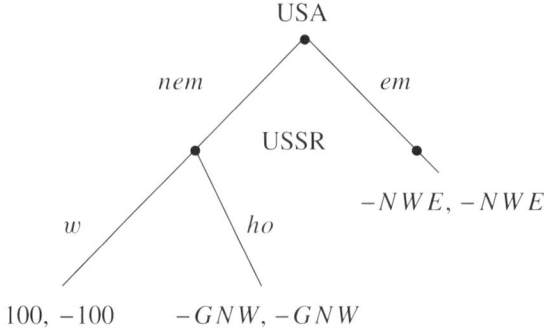

Figure 7.4 The Euromissiles game in extensive form

say, the threat of escalation was used to signal to the USSR that the ultimate responsibility would fall upon the aggressor. This interpretation of nuclear weapons as defensive tools to be used only after a first strike or an invasion extended – at least potentially – even to the strategic component of the arsenal (such as submarines, ICBMs and the bombers controlled by the Strategic Air Command). The US administration decided not to take chances in a risky business and tied its hand, taking a strong commitment to using the strategic arsenal upon aggression, the password being 'burn bridges behind'.

REMARK 7.2   Credible commitments are a possible backbone of solid equilibria. Committing to an all-out nuclear strike by eliminating any possible intermediate step turned out to be a credible measure at the climax of the Cold War.

As with the Cuban missile crisis, also here it can be shown that the equilibrium selected by dominance is also selected by backward induction in the sequential play version of the game, with the USA moving first. The extensive form appears in Figure 7.4.

Choosing strategy *nem*, the White House exploits the first-mover advantage to unmask the Kremlin's bluff, forcing the latter to recognize that withdrawal is better than ceasing to exist. The outcome (*nem, w*) thereby emerges as the subgame perfect equilibrium under perfect information for the USSR. Also, here the presence of multiple Nash equilibria raises an issue as to strategic uncertainty. I leave to you as a useful exercise the task of computing the related probabilities.

7.6   Hawks, doves and Star Wars

Towards the end of the Cold War (in the early 1980s, during the first Reagan administration), the USA started a project known as *SDI* (*Strategic Defense Initiative*). The SDI's objective was to build up a composite defensive shield consisting of a mix of both ground-based and space-based sophisticated weapons to protect the USA from nuclear strikes by enemy ballistic missiles. Owing to its futuristic flavour, SDI became known to the public at large as *Star Wars*.

At first, the project was heavily criticized. Many deemed it completely unrealistic and unfeasible, but it did not peter out completely.[10] During the Clinton administration, it was relabelled *Ballistic Missile Defense Organization*. Subsequently, President George W. Bush revived the project under yet another name, the *National Missile Defense*, announcing the

Matrix 7.6 The benchmark game without
shields

$$B$$

		p	a
A	p	$\pi:\pi$	$-100;\alpha$
	a	$\alpha;-100$	$-50;-50$

deployment of a new ballistic missile defense system in 2002. Some of the sub-programmes embodied in this long list of long-run investment projects have indeed reached operational capabilities, and have been successfully used in theatre operations from the First Gulf War onwards (e.g., the ground-based *Patriot* missile system).

The SDI story lends itself pretty nicely to an analysis based on simple game theory instruments. The following set-up is based on Chassang and Padró i Miquel (2009).

In a nutshell, the issue can be intuitively laid out as follows. Antiballistic shields are inherently defensive instruments – at least at first sight – but can also be considered as dangerously similar to aggressive weapons. The reason is that, say, if country A has a shield while country B hasn't, then the former may conjecture that a nuclear strike on the latter could be carried out successfully as any retaliation by B would be made ineffective by the shield. Of course, B can anticipate this unpleasant consequence of asymmetry and either strongly protest against the construction of a shield by A or build up one of its own (or both).[11]

In line of principle, these alternative scenarios generate three different games. In the first, both countries are endowed with strategic nuclear weapons, but neither country has a shield. In the second, one has built up a shield while the other hasn't. In the third and last game, both countries have a shield. In all cases, they choose between remaining peaceful (p) or attacking (a). The first situation, where shields do not exist, is illustrated in Matrix 7.6.

The key to the interpretation of the game is the size of π relative to the other payoffs. If, as common sense suggests,

$$\pi > \alpha > -50 \tag{7.8}$$

(revealing that all policy-makers are *doves* and value peace more than unilateral aggression), then the matrix portrays a coordination game yielding two Nash equilibria, (p, p) and (a, a), the first being Pareto-efficient. If instead

$$\alpha > \pi > -50 \tag{7.9}$$

the game is a prisoners' dilemma solvable in dominant strategies, with (a, a) as the unique equilibrium.[12] This would be the result of a *hawk*'s attitude, whereby unilateral aggressiveness is better regarded than a peaceful frame of mind.

Now suppose country A builds up a shield s. This increases (respectively, decreases) all payoffs to A (respectively, B) in the case of war, so that the game looks as in Matrix 7.7.

The consequence is that having $\alpha + s > \pi$ becomes *more likely*: the hawk perspective prevailing and resulting in a nuclear holocaust is a concrete risk. Observe that, if $\pi > \alpha$, then *a fortiori* $\pi > \alpha - s$ and therefore B would still contemplate the possibility of remaining peaceful, but this intention is jeopardized by the hawks driving country A to war if indeed $\alpha + s > \pi$, with (a, a) standing out as the unique equilibrium.

As an overall appraisal, let me make one final remark.

Matrix 7.7 Asymmetric Star Wars

$$B$$

		p	a
A	p	$\pi;\pi$	$-100+s;\alpha-s$
	a	$\alpha+s;-100-s$	$-50+s;-50-s$

REMARK 7.3 Star Wars are a risky business. Building up defensive shields is likely to imply that hawks are in control.

The third game, where both countries have their respective shields, can be quickly dealt with under the simplifying assumption that shields neutralize reciprocally and therefore we are back to the same picture as in Matrix 7.6.[13]

To complete the assessment of this problem, look back at Matrix 7.6 and suppose $\pi > \alpha > -50$, so that the game poses a coordination issue. This, of course, prompts the calculation of the mixed-strategy equilibrium. Since the game is symmetric, let's consider the perspective faced by country A, $q \in [0, 1]$ being the probability that country B plays p. Country A is indifferent between p and a if

$$q\pi - (1-q)100 = q\alpha - (1-q)50 \tag{7.10}$$

yielding

$$q^* = \frac{50}{50+\pi-\alpha} \in (0,1) \tag{7.11}$$

and

$$q^* - (1-q^*) = \frac{\pi-\alpha-50}{\alpha-\pi-50} < 0 \quad \text{for all } \pi > \alpha + 50 \tag{7.12}$$

The latter result entails that the probability of bilateral strike is higher than the probability of safeguarding peace even if policy-makers are doves.

Now examine the game from the standpoint of risk dominance (you were asking yourselves why I forced you to go through it in Chapter 3, weren't you?). Evaluating the product of the losses generated by unilateral deviations, we have

$$(\pi-\alpha)(\pi-\alpha) = (\pi-\alpha)^2 \lessgtr 2500 = (-50+100)(-50+100) \tag{7.13}$$

If $\pi - \alpha > 50$, (p, p) risk-dominates (a, a), and conversely. The same conclusion emerges if one takes the risk factors, corresponding to probabilities q^* and $(1-q^*)$, respectively – as we already know from (7.12).[14]

Further reading

More on wargames, at different levels, can be found in Brams (1975), Morrow (1992) and Gardner (1995). The game-theoretic reconstruction of the battle of the Bismarck Sea is in both Luce and Raiffa (1957) and Brams (1975). A formal illustration of wars of attrition and all-pay auctions can be found in Krishna and Morgan (1997). For further views on the MAD game, see Powell (1987), Gardner (1995) and Haas (2001). On arms races and arms control, see Brito and Intriligator (1981, 1994), Lichbach (1990) and Kydd (2000). A detailed analysis of the Euromissiles crisis is given in Smith (1998) and Gaddis (2006), *inter alia*. On the rational use of strategic deterrence, see Lebow and Gross Stein (1989), Powell (1989b, 1990, 2003), Nalebuff (1991), Kilgour and Zagare (1991), Wagner (1991) and Berejikian (2002). The possibility that feeding arms races triggers war has long been debated; see, for instance, Brito and Intriligator (1984, 1996). A fascinating overview of the literature on arms races is presented in Brito and Intriligator (2007). Other very interesting items to read in several related directions are Morrow (1986, 1992), Powell (1989a, 1993) and Tannenwald (1999).

8   Trade, security and hegemony

Commerce with all nations,
alliance with none,
should be our motto.
>   Thomas Jefferson

The topics treated in this chapter are strongly connected with the contents of Chapter 7. In fact, what follows also belongs to the field of international relations. Broadly speaking, one might see it as the complement to the wargames we have just reviewed. This is surely true for:

- the policy discussion concerning openness or international trade, which has generated a lively stream of research in the field of international political economy; and
- the debate about the alternative between guns and butter, investigating the strategy of a country choosing to invest scarce resources either in security or in some other activities enhancing economic welfare but possibly hindering security itself.

Within the theory of international relations, world politics is commonly viewed as *anarchic*, entailing that the emergence of any form of cooperation must be compatible with the principle of sovereignty (see, e.g., Keohane, 1986, p. 1). This, in turn, is compatible with the principle of reciprocity, which lies at the basis of international trade agreements, such as the *General Agreement on Tariffs and Trade* (GATT), and its follow-ups. This simple consideration helps in understanding the links that exist between openness (which may look like a purely economic issue, but is not so), arms races or anything related to security, and the emergence and persistence of hegemony in the international arena. The remainder of this chapter illustrates a few games dealing with this set of issues.

8.1   International cooperation and free trade

I took the Canal Zone
and let Congress debate;
and while the debate goes on,
the canal does also.
>   Theodore Roosevelt

International economics and international political economy share a common interest for a relevant issue, i.e., the choice between protectionism (or autarchy, or the strategic use of tariffs and quotas) on one side, and free trade on the other. However, their respective views

Matrix 8.1 Protectionism versus free trade

$$
\begin{array}{c|cc}
& P & F \\
\hline
1 \quad P & \omega_1(P, P); \omega_2(P, P) & \omega_1(P, F); \omega_2(F, P) \\
F & \omega_1(F, P); \omega_2(P, F) & \omega_1(F, F); \omega_2(F, F)
\end{array}
$$

Above the table is centered: 2

Matrix 8.2 Protectionism as the outcome
of a prisoners' dilemma

$$
\begin{array}{c|cc}
& P & F \\
\hline
1 \quad P & 3; 3 & 12; 1 \\
F & 1; 12 & 8; 8
\end{array}
$$

Above the table is centered: 2

on this issue are quite different, as can be easily ascertained on the basis of the following 2×2 games.

Consider a two-country world, each choosing between *protectionism* (P) and *free trade* (F) in a non-cooperative game under imperfect, symmetric and complete information (see Matrix 8.1). As usual, in every payoff $\omega_i(\cdot, \cdot)$ appearing in the matrix, the first letter indicates the strategy of player i, while the second indicates the strategy of player j, $i, j = 1, 2, i \neq j$.

If $\omega_i(P, P) > \omega_i(F, P)$, $\omega_i(P, F) > \omega_i(F, F)$ and $\omega_i(P, P) < \omega_i(F, F)$, then the game is a prisoners' dilemma, with the free-trade outcome being Pareto-efficient – yet protectionism is a dominant strategy. Replacing letters with appropriate numbers, we may rewrite the matrix as shown in Matrix 8.2.

Here, the free-riding problem affecting every prisoners' dilemma takes the form of an incentive for each country to keep the gates closed and wait for the other to open to trade unilaterally. Obviously, since this holds for both, the unique equilibrium (in strictly dominant strategies) is (P, P), which is Pareto-inefficient as compared to (F, F).

Yet, this is not necessarily the situation that we may expect to arise systematically. Indeed, the so-called *new trade theory* claims exactly the opposite. That is, free trade should arise as a unique and Pareto-efficient equilibrium in strictly dominant strategies (see Helpman and Krugman, 1985; Krugman, 1990; *inter alia*). The explanation of the economic view of international trade and its beneficial effects is based on the idea that consumers are characterized by a love for variety, that is, their utility or satisfaction from consumption increases with the number of varieties of each consumption good they can access. Accordingly, the opening of trade (i) fosters price competition among firms on the world market, and (ii) expands the choice set for all consumers. Both factors operate in the direction of a welfare increase.

What changes as compared to the previous case is the list of inequalities, $\omega_i(P, P) < \omega_i(F, P)$ and $\omega_i(P, F) < \omega_i(F, F)$. Hence, the game now looks like in Matrix 8.3.

Matrix 8.3 Free trade as a Pareto-efficient
equilibrium

$$
\begin{array}{c|cc}
& P & F \\
\hline
1 \quad P & 3; 3 & 4; 5 \\
F & 5; 4 & 8; 8
\end{array}
$$

Above the table is centered: 2

The two basic factors behind this different approach to free trade are (i) the preference for variety, which is assumed to dwell in the representative consumer's mind, as long as countries are sufficiently symmetric (i.e., similar in terms of income and productivity, for instance), and (ii) the assumption of monopolistic competition (largely equivalent to perfect competition except for the fact that firms supply differentiated products).[1] On the basis of these assumptions, unilateral opening to trade leads in any case to an increase in home consumers' surplus – hence the policy-maker of each country has a strict incentive to open.

Clearly, whether the first or the second perspective is more attractive depends on the ideological background of the scholar examining the matter, in combination with empirical observations. Both approaches have their own unquestionable merits, as well as some drawbacks.

Be that as it may, it is possible to show that, even if the underlying game is a prisoners' dilemma, one can reasonably hope to observe free trade in equilibrium. This is a typical task for the theory of repeated games that we met in Chapter 5. Provided the game is repeated over an infinite horizon, the so-called folk theorem (in any of its manifold formulations) ensures that players may play a subgame perfect equilibrium Pareto-dominating that of the constituent game.

Here we shall adopt the rules of the perfect folk theorem (Friedman, 1971), under the assumption that Matrix 8.1 be a prisoners' dilemma. Let me recall briefly the structure of the supergame on which Friedman's result is based. Time is $t = 0, 1, 2, 3, \ldots, \infty$, and both countries discount future gains using the same discount factor $\delta \in [0, 1]$. Remember that, if $\delta = 0$, countries are completely myopic – accordingly, they are altogether unaware that the game is a repeated one; conversely, the higher is δ, the higher is the value attached to future payoffs. That is, players become progressively more forward-looking as δ increases.

The rules of the repeated game are based on the following grim trigger strategies:

1   At $t = 0$, play F.
2   At any $t > 0$, play F if both have played F at $t - 1$. Otherwise, play P and keep on like that forever.

Strategy F is part of a subgame perfect equilibrium provided that

$$\sum_{t=0}^{\infty} \omega_i(F, F)\delta^t \geq \omega_i(P, F) + \sum_{t=1}^{\infty} \omega_i(P, P)\delta^t \tag{8.1}$$

In (8.1), the expression on the left-hand side is the discounted flow of payoffs (assessed at $t = 0$) generated by sticking to free trade forever. The expression on the right-hand side is the discounted flow of payoffs generated by the unilateral deviation towards protectionism, followed by the Nash punishment. Since

$$\sum_{t=0}^{\infty} \delta^t = \frac{1}{1 - \delta} \qquad \Leftrightarrow$$

$$\sum_{t=1}^{\infty} \delta^t = \sum_{t=0}^{\infty} \delta^t - 1 = \frac{1}{1 - \delta} - 1 = \frac{\delta}{1 - \delta} \tag{8.2}$$

Matrix 8.4 Here, a 'global power' has a
taste for free trade . . .

		France	
		P	F
GB	P	1; 3	2; 1
	F	3; 4	4; 2

condition (8.1) can be reformulated as

$$\frac{\omega_i(F, F)}{1 - \delta} \geq \omega_i(P, F) + \frac{\delta}{1 - \delta} \omega_i(P, P) \tag{8.3}$$

yielding

$$\delta \geq \frac{\omega_i(P, F) - \omega_i(F, F)}{\omega_i(P, F) - \omega_i(P, P)} = \delta^* \tag{8.4}$$

which represents the stability condition that both countries must satisfy so that the repeated game may indeed yield (F, F) as the subgame perfect equilibrium. Otherwise, if $\delta \in [0, \delta^*)$, then the equilibrium outcome is (P, P) at any $t = 0, 1, 2, 3, \ldots, \infty$. In the latter case, the countries are unable to escape from the prisoners' dilemma.

Now take the numerical payoffs appearing in Matrix 8.2 and plug them into (8.4). What you get is

$$\delta \geq \delta^* = \frac{12 - 8}{12 - 3} = \frac{4}{9} \tag{8.5}$$

establishing that, in this example, free trade is sustainable as the long-run equilibrium for all $\delta \geq 0.\bar{4}$ (the bar indicates that this is a periodical number, whereby $0.\bar{4} = 0.44444\ldots$).

This discussion can be summarized by stating the following remark.

REMARK 8.1 If countries are sufficiently patient (i.e., if they attach a sufficiently high weight to future payoffs), then repeating the constituent game over an infinite horizon allows them to sustain free trade as an equilibrium outcome forever.

What if countries are not symmetric? Matrix 8.4 summarizes an asymmetric game that took place in the nineteenth century between Great Britain and France. Great Britain, after the outcomes of the Seven Years' War and the Napoleonic wars – and ultimately thanks to the skills and bravery of the Royal Navy – was enjoying a hegemonic position over the globe.[2] That was probably the earliest example ever of a 'global power'.

This game, described by Keohane (1986, p. 15) and Gowa (1989, p. 1251), has a unique equilibrium, in strictly dominant strategies: (F, P), with Great Britain opening to trade and France keeping a protectionist stance.

This example helps to clarify that incentives towards trade may differ significantly across countries. In slightly different terms, we may pass on to examine a game between a 'large' or *hegemonic* country (H), and a 'small' or *non-hegemonic* one (NH). Matrix 8.5, describing this situation, is taken from Conybeare (1984).

In this case, the small country has a strictly dominant strategy, F, while the big country hasn't any. However, by iterating the dominance criterion, once NH has dropped the first

Matrix 8.5 ... but the reverse may also
apply

		NH	
		P	*F*
H	*P*	1; 1	4; 2
	F	2; 3	3; 4

column, what remains of the first row identifies in P the dominant strategy for H. Therefore, (P, F) emerges as the unique Nash equilibrium in dominant strategies. This equilibrium is also selected by backward induction, provided that H is the first mover in the perfect information game. Note, however, that, if the small country plays first, the subgame perfect information is (F, P) – which isn't even a Nash equilibrium if we assess the above matrix under imperfect information.[3]

8.2 Guns versus butter and the trade-off between openness and security

> War is a game that is played with a smile.
> If you can't smile, grin.
> If you can't grin, keep out of the way till you can.
>
> Sir Winston Churchill

In its starkest form, the realistic approach to international relations views them as an anarchic system where nations allocate scarce resources to alternative objectives, such as security on one side and economic growth on the other. This is the essence of the *guns versus butter trade-off* that we will investigate in this section.

Possibly the easiest way to sketch the problem of choosing between guns and butter is the game represented in Matrix 8.6. Here, two identical countries, 1 and 2, choose between two mutually exclusive strategies, G (guns) and B (butter). Let π be the payoff each country gets if this game were not played at all. If both go for the guns, the countries' arsenals balance off reciprocally, so that the resulting individual payoff is $\pi + g - g = \pi$. If both choose butter, the individual payoff is increased by the amount b. In the asymmetric outcomes, the country that plays B suffers from the fact that the other is building up weapons.

The equilibrium of the game depends upon the relative size of b and g, i.e., the countries' preferences between guns and butter:

- If butter matters more than guns ($b > g$), then $\pi + b - g > \pi$ and $\pi + b > \pi + g > \pi$. Consequently, the game is solvable by strict dominance, with (B, B) emerging as the unique and Pareto-efficient equilibrium.
- If, instead, guns matter more than butter ($g > b$), then $\pi + b - g < \pi$ and $\pi + g > \pi + b > \pi$. The game is again solvable in dominant strategies, but this time the

Matrix 8.6 Guns versus butter

		2	
		G	*B*
1	*G*	$\pi; \pi$	$\pi + g; \pi + b - g$
	B	$\pi + b - g; \pi + g$	$\pi + b; \pi + b$

Matrix 8.7 Guns versus butter with asymmetric
preferences

		2	
		G	B
1	G	$\pi + g_1 - g_2; \pi + g_2 - g_1$	$\pi + g_1; \pi + b_2 - g_1$
	B	$\pi + b_1 - g_2; \pi + g_2$	$\pi + b_1; \pi + b_2$

unique equilibrium is (G, G), which is Pareto-dominated by (B, B). Therefore, in this case the game is a prisoners' dilemma.

So far, I have taken the two countries to be fully symmetric, but this is not necessarily realistic, as history and casual observation alike immediately suggest: there are countries that attach a higher weight to security (or guns), while others do the opposite. A slightly more general version of the game, described by Matrix 8.7, captures this aspect and allows one to grasp the possible consequences of this particular asymmetry in tastes.

Assume country 1 prefers butter to guns ($b_1 > g_1$), while country 2 prefers guns to butter ($g_2 > b_2$). This implies the following inequalities:

$$\pi + b_1 - g_2 > \pi + g_1 - g_2, \quad \pi + b_1 > \pi + g_1 \tag{8.6}$$

$$\pi + g_2 - g_1 > \pi + b_2 - g_1, \quad \pi + g_2 > \pi + b_2 \tag{8.7}$$

Accordingly, B is dominant for country 1 while G is dominant for country 2. The resulting equilibrium, (B, G), is also unique.

A more detailed approach to the same issue can be framed in the following terms. Historical facts suggest that trade restrictions (tariffs, quotas, embargoes and other sanctions) have often been adopted as part of a security policy. The restrictions imposed by the USA on Japan relating to oil and raw materials on the eve of the war in the Pacific are just one example. The following game, loosely based on Skaperdas and Syropoulos (2001), illustrates this problem.

Players are countries 1 and 2. Each of them may open to trade (strategy t) or build up armaments (strategy a) to exert some degree of control over a strategic natural resource r (oil, uranium, etc.) located in a third country. Let the cost of manufacturing weapons be k. The layout of the game is shown in Matrix 8.8.

The two countries are fully symmetric. Under bilateral trade (outcome (t, t)), each country imports and exports the same quantity of goods x, so that the payoff in this regime is the level of consumption associated with domestic production, c, plus the balance between imports and exports, which is nil: $c + x - x = c$. The other symmetric outcome is that where frontiers are closed and countries invest in an arms race to control the natural resource. In such a case, assume for simplicity that each country ends up acquiring 50 per cent of such resource. The associated payoff is the sum of the level of autarchic consumption plus half

Matrix 8.8

		2	
		t	a
1	t	$c; c$	$c - x - r; c + x + r - k$
	a	$c + x + r - k; c - x - r$	$c + r/2 - k; c + r/2 - k$

the natural resource endowment, minus the investment cost: $c + r/2 - k$. The asymmetric outcome yields complete control of the resource to one country, while the other is put at a disadvantage for the very same reason. This adds up to the asymmetry in trade flows, as the aggressive country also exports to the non-aggressive one.

In order to assess the countries' incentives as well as the equilibrium outcome(s), we have to make some reasonable guess about the relative sizes of the parameters. Intuitively, it must be that $r > k$, i.e., the costly acquisition of the full control over the resource must be convenient. However, $r/2$ can be either higher or lower than k. With this in mind, observe that

$$c + x + r - k > c \tag{8.8}$$

always, while

$$c + \frac{r}{2} - k > c - x - r \tag{8.9}$$

is equivalent to

$$\frac{3r}{2} - k > -x \tag{8.10}$$

which is also true because we are assuming $r > k$. As a result, trade is a dominated strategy and will not be adopted. The unique equilibrium, at the intersection of strictly dominant strategies, is (a, a). Whether this identifies a prisoners' dilemma depends on the sign of $r/2 - k$. If this is a positive number, then the equilibrium is Pareto-efficient; otherwise the game is indeed a prisoners' dilemma. It is worth stressing that the structure of countries' preferences may prevent them from considering the game as a prisoners' dilemma, and therefore they may view the perspective of going for guns as not only rational but also *efficient* or even *desirable*.

Accordingly, the bottom line of the foregoing discussion can be expressed as follows.

REMARK 8.2   The trade-off between guns and butter can be expected to drive countries in the direction of guns. Moreover, they may not see it as the outcome of a prisoners' dilemma.

8.3   The persistence of unipolarism

> Concentrated power has always
> been the enemy of liberty.
>> Ronald Reagan

This section takes a look at the problem of hegemony as a unilateral phenomenon.[4] After the end of the Cold War, the USA remained the only global power. This situation has been labelled as *unipolarism*. However, according to the realist theory, unipolarism is not bound to be a long-run equilibrium.[5] Consequently, the current *status quo* seems at odds with the dominant view of the theory of the balance of power, establishing that sooner or later a counterbalancing power, perhaps taking the form of a coalition of several states, should arise. This position is largely based on history, and therefore sounds like a robust one, but is there no space at all for exceptions?

Indeed, in contrast with this position, Ikenberry (2002)[6] proposes an eclectic alternative view of the issue at stake, whereby the hegemony of the global power is sustained by an international governance system granting every potential competitor a number of appealing benefits, e.g., bilateral trade agreements, military support, technology transfers, etc.

This mixture of liberal and realist attitudes has two major implications. On the one hand, it limits the power effectively enjoyed by the USA as the leader of the international order; on the other, it clearly tends towards a reduction of the incentives to look for alternatives (e.g., a counterbalancing coalition) by means of a growing network of international institutions. This amounts to saying that the system introduced by the USA after the end of World War II contains a trade-off between the amount of effective hegemony exerted by the global power and the degree of confidence that the *status quo* might indeed persist.[7] That is, the insurance against defections is costly. This fact has prompted several doubts on the robustness of the existing system (see, e.g., Kupchan, 2002), as the sustainability of such costs on the part of the USA (and their taxpayers) is not completely out of the question.

The following is a three-country world[8] where the players are a global power and two satellites. To keep things as essential as possible, the strategy space is binary, both for the global power and for each of the satellites. The hegemonic country has to choose whether or not to subsidize the satellites, so as to avoid the formation of a counterbalancing coalition, while the satellites have to decide whether to ally or not, being aware that constructing a coalition is a costly activity.

8.3.1   The model

The players are one global power (denoted by G) and two satellites, s_1 and s_2. Assume that: (i) G may pay each of the satellites a given subsidy $\sigma > 0$, lest they get together to form a coalition aimed at diminishing G's hegemonic position; and (ii) each satellite may, if it does not receive the subsidy, invest an amount of resources $k > 0$ to build up the coalition. It is relevant to stress that the alliance is taken not to be a pure public good, in that an amount equal to $2k$ is needed for its construction. To clarify this aspect, I introduce the following assumption.

ASSUMPTION The overall cost of building up a coalition is $2k$. The amount of resources available to each satellite is k.

This assumption prevents each small country s_i individually undertaking the venture of building up a countervailing power to counterbalance that of G. In turn, this might open a discussion as to whether G could save upon the total amount of subsidies by granting only a single σ to one of the satellites, obtaining in return that the coalition does not arise. Here I will not consider this possibility, the reason being twofold. First, the distribution of subsidies to both satellites is in accordance with facts, as observing the behaviour of the USA towards, say, Russia and China suffices to confirm. Second, the model is defined in such a way that it includes only the relevant players, where *relevant* means that they are the only potential competitors that G must take into account. The rest of the world (say, Japan, the EU, etc.), which is known not to represent a threat to the global power in this respect, is not modelled.

I am now in a position to introduce the basic elements of the game. The global power faces four alternative perspectives:

1   If it delivers the subsidies and the satellites do not build up a coalition,[9] G's utility is

$$U_G(\sigma, NC) = c_G - 2\sigma + 2\beta\sigma + P \tag{8.11}$$

where:

- *NC* indicates that there is no coalition;
- c_G is G's domestic consumption level;
- σ is the size of the subsidy, and therefore 2σ is the total cost borne by G;
- $2\beta\sigma$ measures the benefit to the global power, generated by the subsidies to the satellites, the marginal benefit being measured by parameter $\beta \in (0, 1)$; and
- P measures the power level enjoyed by G if no coalition builds up, and $U_G(\sigma) > 0$ for all $\sigma \in (0, (c_G + P)/[2(1 - \beta)])$.

The relevance of β, to which I will come back later in the analysis of the game, can be preliminarily illustrated on the basis of two alternative perspectives. Keeping the *status quo* unchanged through the distribution of subsidies implies a cost and a return to G. The first is obviously represented by the total amount of subsidies, i.e., the quantity -2σ; the second is a positive component measuring the valuation attached by G to the fact that satellites do not ally against it, $2\beta\sigma$. Hence, clearly we must have $\beta < 1$ for the subsidization manoeuvre to entail a net cost, as common sense would suggest. As is going to become apparent in the remainder of the book, the size of β will play a crucial role in shaping the equilibrium outcome.

2 If, instead, the global power does not deliver any subsidies to the satellite countries, and the latter form a coalition C, then G's payoff is

$$U_G(0, C) = c_G + P - \beta C \tag{8.12}$$

where $C \in (0, P)$ measures the level of power held by the alliance between satellites s_1 and s_2. Note that the coalition reduces G's power by an extent measured by parameter β. This amounts to saying that G evaluates in the same way the marginal disutility of any power decrease as well as that associated to paying subsidies.[10]

3 The third situation is that where G pays the subsidies but the satellites decide to invest in a counterbalancing coalition, yielding thus

$$U_G(\sigma, C) = c_G - 2\sigma + P - \beta C \tag{8.13}$$

4 The last possible case is that where the global power does not distribute the subsidies and, this notwithstanding, the satellites do not form an alliance. In this situation, G's utility is

$$U_G(0, NC) = c_G + P \tag{8.14}$$

Now focus upon the satellites. Again, four cases can be envisaged:

1′ If they accept the subsidies and do not create a coalition, each satellite attains the following utility:

$$u_i(\sigma, NC) = c_i + \sigma \tag{8.15}$$

where c_i defines the domestic consumption level in satellite i.

2′ If, instead, having received no subsidy, the satellites build a coalition, their individual utility becomes

$$u_i(0, C) = c_i - k + C \tag{8.16}$$

Matrix 8.9 Will unipolarism persist?

		s_i	
		NC	C
G	σ	$c_G - 2\sigma(1-\beta) + P; c_i + \sigma$	$c_G - 2\sigma + P - \beta C; c_i + \sigma + C - k$
	a	$c_G + P; c_i$	$c_G + P - \beta C; c_i + C - k$

In this regard, it is worth stressing that country i pays an individual fee k associated with the investment required to construct the alliance, but then enjoys the full benefit C yielded by the alliance itself. One has to assume that the net payoff generated by the coalition is positive, $C \geq 2k$. This suffices to ensure that $u_i(0, C) > 0$.

3′ In this case, each of the satellites is subsidized but still they decide to form a coalition, whereby the individual payoff accruing to each of them is

$$u_i(\sigma, C) = c_i + \sigma + C - k \qquad (8.17)$$

4′ Finally, there remains the case where G does not distribute any subsidy and the satellites do not invest to build an alliance:

$$u_i(0, NC) = c_i \qquad (8.18)$$

The game can be represented in strategic form as in Matrix 8.9, where G is the row player. Given the additive separability of the payoff functions, I will only represent the payoffs of one of the two satellites as the column player, without loss of generality and without affecting the solution.[11]

Assume that the game is one of imperfect, complete and symmetric information. In order to characterize the equilibrium solution, one has to examine the incentives, respectively, for G to pay the subsidies so as to try to avoid the arising of the coalition, and for the satellites to accept the subsidies and, possibly, not to form the coalition. Observe that

$$U_G(\sigma, NC) < U_G(0, NC) \quad \Leftrightarrow \quad \beta \in (0, 1)$$
$$U_G(\sigma, C) < U_G(0, C) \quad \Leftrightarrow \quad \sigma > 0 \qquad (8.19)$$

that is, (i) if the satellites do not form a coalition, then it is convenient for the global power not to pay the subsidies for all the admissible values of β, while (ii) in presence of the coalition, it is optimal for G not to distribute any subsidy for all admissible $\sigma > 0$. Therefore, the following remark can be made.

REMARK 8.3 The global power has a strictly dominant strategy, which consists in paying no subsidies to the satellites, for all admissible levels of parameters β and σ.

As for the satellite country i, the following hold:

$$u_i(\sigma, C) > u_i(\sigma, NC) \quad \Leftrightarrow \quad C > k$$
$$u_i(0, C) > u_i(0, NC) \quad \Leftrightarrow \quad C > k \qquad (8.20)$$

which entails the following remark.

REMARK 8.4 Forming a coalition is a strictly dominant strategy for the satellites.

On the basis of Remarks 8.3 and 8.4, without further proof, I may state the following result.

PROPOSITION 8.1 *The game has a unique Nash equilibrium identified by the strategy pair* $(0, C)$. *Such a Nash equilibrium is in strictly dominant strategies.*

The next step consists in verifying whether the present game can be a prisoners' dilemma. For this to hold, the additional condition that has to be satisfied is that the equilibrium be Pareto-inefficient, which depends on the relative magnitude of the payoffs appearing along the main diagonal of Matrix 8.9.

If the following inequalities are simultaneously satisfied, then the game is indeed a prisoners' dilemma:

$$U_G(\sigma, NC) > U_G(0, C) \quad \Leftrightarrow \quad c_G - 2\sigma(1 - \beta) + P > c_G + P - \beta C \tag{8.21}$$

$$u_i(\sigma, NC) > u_i(0, C) \quad \Leftrightarrow \quad c_i + \sigma > c_i - k + C \tag{8.22}$$

Simplifying condition (8.21), one obtains

$$\sigma < \frac{\beta C}{2(1 - \beta)} \tag{8.23}$$

while condition (8.22) yields

$$\sigma > C - k \tag{8.24}$$

An intuitive interpretation can be given for both. On the one hand, (8.23) shows that G will find it profitable to deliver the subsidy if it is low enough as compared to the ratio between the damage generated by the coalition, given by βC, and the net marginal benefit of the subsidies, measured by $2(1 - \beta)$. On the other hand, condition (8.24) simply states that a satellite will be willing to accept the subsidy and abandon the perspective of investing to build up the coalition if the subsidy is higher than the net gain associated with building up the coalition.

Note that the compatibility between (8.23) and (8.24) requires that

$$\frac{\beta C}{2(1 - \beta)} > C - k \tag{8.25}$$

which is equivalent to the following condition:

$$2k(1 - \beta) > C(2 - 3\beta) \tag{8.26}$$

Whether inequality (8.26) is met will of course depend on the relative size of the three parameters involved. However, there clearly exist admissible regions of parameter space where this condition holds. The validity of this assertion is quickly shown, as follows. By assumption, we know that $1 - \beta > 0$. Therefore, the left-hand side of (8.26) is positive. The right-hand side may take either sign, depending on the value of β: it is positive for all $\beta \in (0, 2/3)$, while it is negative for all $\beta \in (2/3, 1)$. Therefore, taking any values of β in

(2/3, 1) is sufficient (but not necessary) to validate (8.26).[12] To complete the argument, it suffices to observe that, if (8.26) holds, G may always choose an appropriate value of σ such that the resulting game is a prisoners' dilemma. Hence, I have proved the following result.

PROPOSITION 8.2   *Examine the size of the subsidy. Two mutually exclusive situations may arise:*

- *Suppose that $2k(1 - \beta) > \max\{0, C(2 - 3\beta)\}$. Then $C - k > \beta C/[2(1 - \beta)]$, and for all $\sigma \in \big(C - k, \beta C/[2(1 - \beta)]\big)$ the Nash equilibrium in dominant strategies identified by $(0, C)$ is the outcome of a prisoners' dilemma. If either $\sigma > \beta C/[2(1 - \beta)]$ or $\sigma \in (0, C - k)$, the outcomes (σ, NC) and $(0, C)$ cannot be Pareto-ranked.*
- *Suppose instead that $C(2 - 3\beta) > 2k(1 - \beta) > 0$. Then $C - k < \beta C/[2(1 - \beta)]$, and for all $\sigma \in \big(\beta C/[2(1 - \beta)], C - k\big)$ the Nash equilibrium in dominant strategies identified by $(0, C)$ is Pareto-efficient. If either $\sigma > C - k$ or $\sigma \in \big(0, \beta C/[2(1 - \beta)]\big)$, the outcomes (σ, NC) and $(0, C)$ cannot be Pareto-ranked.*

The second part of the above proposition describes the case where the obvious solution is one where satellites do ally, the global power does not pay any subsidies to try to keep the *status quo* unaltered, and ultimately there are no regrets on either side. More interesting is the alternative perspective where we observe a prisoners' dilemma game, because in such a case, as is well known, a repeated game framework may offer a way out of the Pareto-inefficient Nash equilibrium generated by the one-shot game.[13] This is done in the next section.

8.3.2   *The supergame*

To model the behaviour of players in the supergame, I will resort here to the so-called perfect folk theorem based upon grim trigger strategies (Friedman, 1971).[14]
   Assume that

$$\frac{C}{2k} > \frac{\beta - 1}{3\beta - 2} \quad \text{and} \quad \sigma \in \left(\frac{\beta - 1}{3\beta - 2}, \frac{C}{2k}\right)$$

so that the constituent game described by Matrix 8.9 is indeed a prisoners' dilemma. It is important to stress that G may intentionally choose a level of σ in this interval, in order to make the stage game a prisoners' dilemma. The relevance of this aspect lies in the fact that, by doing so, the global power may deliberately induce the development of a supergame that may have an equilibrium outcome that differs completely from the one characterizing the one-shot game. It is also worth emphasizing that only G is able to manipulate the strategic incentives underlying the game, while the satellites cannot do so. To some extent, this says that one of the prerogatives of the global power is precisely to affect at will the nature of the game, if this appears to be profitable for the global power.
   Let the stage game repeat over discrete time $t = 0, 1, 2, 3, \ldots, \infty$.[15] Players share the same discount factor $\delta \in [0, 1]$. According to the rules of the perfect folk theorem, it is established that players stick to the Pareto-efficient path as long as no deviation is detected. As soon as this happens, then they revert to the dominant Nash equilibrium strategy forever. Therefore, the strategy pair (σ, NC) is a subgame perfect equilibrium of the supergame if and only if, for both players, the discounted flow of payoffs associated with the collusive outcome (σ, NC) is at least as large as the discounted flow of payoffs associated with the

alternative perspective where either player defects once and then both revert to the one-shot non-cooperative equilibrium forever. That is, the outcome (σ, NC) is sustainable if and only if the following inequalities are both satisfied:

$$\frac{U_G(\sigma, NC)}{1-\delta} \geq U_G(0, NC) + \frac{\delta U_G(0, C)}{1-\delta} \tag{8.27}$$

$$\frac{u_i(\sigma, NC)}{1-\delta} \geq u_i(\sigma, C) + \frac{\delta u_i(0, C)}{1-\delta} \tag{8.28}$$

That is, unlike the previous supergames we have seen, the asymmetry across players characterizing the present game complicates the analysis a little bit, as the global power and the satellites have different views about the long-run perspective they are collectively facing.

From (8.27), one obtains

$$\delta \geq \frac{2(1-\beta)\sigma}{\beta C} \equiv \delta_G \tag{8.29}$$

while (8.28) yields

$$\delta \geq \frac{C-k}{\sigma} \equiv \delta_s \tag{8.30}$$

with $\delta_G, \delta_s \in (0, 1)$ on the basis of conditions (8.23) and (8.24). A straightforward examination of δ_G and δ_s reveals what is stated in the following remark.

REMARK 8.5 The discount factor δ_G increases monotonically in σ, while the opposite holds for the discount factor δ_s.

To prove this result, it suffices to observe that σ appears in the numerator of δ_G with positive sign (as $\beta < 1$), whereas it appears in the denominator of δ_s (again with positive sign). The intuitive explanation of these facts is that, as σ becomes larger, it becomes increasingly costly for G to pay the subsidies, while it becomes more attractive to each of the satellites to accept the subsidies and to renounce building the coalition.

Moreover, we find that

$$\delta_G - \delta_s \propto 2(1-\beta)\sigma^2 - \beta(C-k)C \tag{8.31}$$

which is positive if and only if

$$\sigma > \sqrt{\frac{\beta(C-k)C}{1(1-\beta)}} \tag{8.32}$$

Comparing the expression on the right-hand side of (8.32) with the lower and upper bounds of the interval for σ wherein we observe a prisoners' dilemma, we find that

$$\sqrt{\frac{\beta(C-k)C}{1(1-\beta)}} \in \left(C-k, \frac{\beta C}{2(1-\beta)} \right) \tag{8.33}$$

provided that (8.26) is satisfied. This allows me to formulate the main result of this chapter.

THEOREM 8.1 *Take* $2k(1-\beta) > \max\{0, C(2-3\beta)\}$, *so that the game as defined by Matrix 8.9 is a prisoners' dilemma. Then, consider the supergame generated by the infinite repetition of such a constituent game. Using the perfect folk theorem, the resulting critical thresholds of the discount factors are*

$$1 > \delta_G > \delta_s > 0 \quad for\ all\ \sigma \in \left(\sqrt{\frac{\beta(C-k)C}{1(1-\beta)}}, \frac{\beta C}{2(1-\beta)} \right)$$

$$1 > \delta_s > \delta_G > 0 \quad for\ all\ \sigma \in \left(C-k, \sqrt{\frac{\beta(C-k)C}{1(1-\beta)}} \right)$$

If $\delta \geq \max\{\delta_G, \delta_s\}$, *then the strategy pair* (σ, NC) *is sustainable at the subgame perfect equilibrium of the repeated game. For all* $\delta \in [0, \max\{\delta_G, \delta_s\})$, *the subgame perfect equilibrium is* $(0, C)$.

It is worth observing that, for comparatively low levels of the subsidy, the most demanding threshold is associated with the time preferences of the global power; conversely, for sufficiently high values of σ, the decisive threshold is δ_s. A possible interpretation of this outcome could be that, if σ is large enough, then the prospect of cheating becomes very appealing for each of the satellites. By doing so, each of them receives, if only for one period, a very rewarding subsidy without bearing the cost of an alliance.

As a last remark, observe that the repeated game still allows for the Nash equilibrium of the constituent game to repeat forever, that is, the above theorem illustrates, as is always the case with repeated games, the existence of multiple equilibria, one of which is that yielded by the prisoners' dilemma. However, the point made by the present analysis is that the persistence of a *status quo* characterized by unipolarism cannot be rationally ruled out.

Further reading

For more on the issues tackled in this chapter, see Brams (1975), Waltz (1979), Gilpin (1981), Keohane (1989) and Ikenberry (2001, 2002). On trade theory and policy, see Helpman and Krugman (1989), Krugman (1991) and Grossman (1992). On the interplay between security and openness (or trade), see Dorussen (1999). On the trade-off between guns and butter, see Polachek (1980), Chan and Mintz (1992), Powell (1993), Gowa (1994), Heo (1998) and Carrubba and Singh (2004). Concerning the theory of the balance of power, unipolarism and hegemony, see Walt (1987), Mastanduno (1997), Vasquez and Elman (2000), Ikenberry (2004) and Cox (2005).

8.4 Appendix: the game between satellites

Here we may examine the strategic interaction between satellites, which have to choose whether or not to undertake the construction of the coalition. To do so, let's take a fully non-cooperative perspective. Two alternative games can be envisaged, depending on whether or not the global power subsidizes the satellites. Matrix 8.10 describes the first case.

Clearly, the game is symmetric along the main diagonal and there is no dominant strategy, with

$$\begin{aligned} c_i + \sigma &> c_i + \sigma - k \\ c_i + \sigma + C - k &> c_i + \sigma \end{aligned} \tag{8.34}$$

Matrix 8.10 Satellites receive subsidies

		S_2	
		0	k
s_1	0	$c_1 + \sigma; c_2 + \sigma$	$c_1 + \sigma; c_2 + \sigma - k$
	k	$c_1 + \sigma - k; c_2 + \sigma$	$c_1 + \sigma + C - k; c_2 + \sigma + C - k$

Therefore, matrix 1 describes a coordination game with two symmetric equilibria, $(0, 0)$ and (k, k), which can be Pareto-ranked: $(k, k) \succ (0, 0)$, as both countries strictly prefer to ally. However, if the game is played under imperfect information, i.e., simultaneously, then there is no particular reason to believe that the satellites will select (k, k).[16] Completely different considerations can be put forward under perfect information, i.e., if the game is played sequentially with either player at the root of the game tree. The extensive form is shown in Figure 8.1.

It is easily checked by backward induction that the subgame perfect equilibrium is (k, k). To see this, examine each subgame in isolation, starting from the terminal nodes. If satellite j knows itself to be in the left singleton (because i has chosen strategy 0), then the optimal choice is to play 0. The resulting payoffs are $c_i + \sigma$ for both.

Otherwise, corresponding to the right singleton, j's best reply to k consists in choosing k, the associated payoffs being $c_i + \sigma + C - k$. Hence, i is aware that choosing 0 it will attain $c_1 + \sigma$, while choosing k it will attain $c_1 + \sigma + C - k > c_i + \sigma$. As a result, the optimal choice for the player located at the root of the game tree is to invest k, inducing an analogous behaviour by the country located at the intermediate nodes.

The alternative case where G does not pay any subsidies can be quickly dealt with, as the game is qualitatively equivalent except for the absence of σ in all of the payoffs involved, as clarified by Matrix 8.11.

Therefore, under imperfect information, we again have two Nash equilibria in undominated strategies, $(0, 0)$ and (k, k), the latter being selected as the unique subgame perfect equilibrium by backward induction, if the game is solved under perfect information.

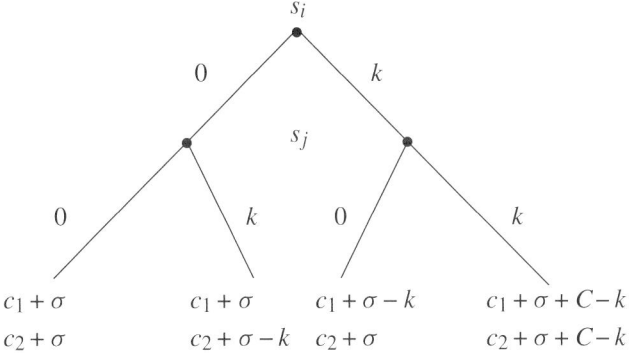

Figure 8.1 The extensive form of Matrix 8.10

Matrix 8.11 Satellites do not receive subsidies

		S_2	
		0	k
s_1	0	$c_1; c_2$	$c_1; c_2 - k$
	k	$c_1 - k; c_2$	$c_1 + C - k; c_2 + C - k$

9   The role of information

A good marriage would be between
a blind wife and a deaf husband.

Honore de Balzac

I don't necessarily agree with Balzac, but his words are in line with the topics treated in this chapter. So far, we have been dealing with games where information was symmetric and complete, and with either perfect or imperfect information. It is now time to go into the realm of asymmetric and/or incomplete information games.

First of all, let me recall the notions of information symmetry and completeness spelled out in Chapter 2. Information is complete provided that each player knows the entire structure of the game, i.e., the identities of all players, the admissible strategies for each of them, and their respective payoffs corresponding to every admissible outcome. Information is symmetric provided that exactly the same amount of information is available to all players. If any of these stipulations does not apply, then the resulting game is affected by either incomplete or asymmetric information, or both.

It goes without saying that the assumptions of symmetry and completeness are, in many situations, much less than realistic, if at all so. Therefore, being able to cope with informational problems has been a primary goal in the agenda of game theorists for decades, and the related literature is still flourishing. The solution of such games requires new notions about the way players think of themselves and rivals before and during the game, and ultimately asks for new equilibrium concepts elaborating the limited information players may rely upon *ex ante*, as well as the additional bits of information they avidly extract during the development of the game.

As we shall see, the presence of uninformed players may create a temptation for informed ones to manipulate information strategically, by sending signals that may be either true or false, depending on the senders' incentives. This aspect is also intertwined with (and, to a significant extent, conditioned by) the endogenous construction of reputation by informed players, on which uninformed ones can rely, and that contributes to shape the outcome of a game.

I will start out by showing you the basic elements of asymmetric information. Then, I will pass on to games taking place under incomplete information. Finally, we'll have a look at a solution concept that may be used also when asymmetric or incomplete information is not an issue: forward induction. An aspect that so largely contributes to make forward induction as interesting as it is for game theorists is its relation (or contrast) with the backward induction method you are well acquainted with.

9.1 Asymmetric information

The usual approach to modelling the presence of asymmetric information in games consists in viewing the latter as situations where agents have to stipulate agreements concerning a certain transaction or their future behaviour. That is, the heart of the matter is a contract regarding any objects being traded or relevant aspects of the future relationship between these agents in the future.

Asymmetric information can take two forms, alternatively. If asymmetry precedes the transaction (or the stipulation of the contract), we observe a problem of *adverse selection*; if instead asymmetric information arises after the stipulation of the contract, we are in a condition of *moral hazard*.

For the sake of simplicity, let's confine our attention to two-player games. Under asymmetric information, the agent endowed with a larger amount of information is called the *agent*, while the other is called the *principal*. Their interplay defines what is traditionally labelled as an *agency relationship*.

According to Stiglitz (1985), adverse selection problems regard the features of objects being traded (as, for example, the built-in quality of consumption goods) that are supposed to be known *ex ante* to sellers but not necessarily to purchasers, while the phenomenon of moral hazard is connected with the behaviour of agents (as, for example, workers shirking once hired). There are, of course, hybrid cases. One obvious example is the insurance markets, where the insurance company (the principal) cannot tell *a priori* whether any specific client (the agent) buying a theft insurance for his/her car is a risky individual or not: he/she may usually leave the car carelessly unlocked in a parking lot, thus increasing the probability of theft. This situation envisages the presence of an *ex ante* adverse selection issue about the type of the client, as well as an *ex post* moral hazard issue once the insurance has been activated.

Within the moral hazard area (following Arrow, 1985), we can build up an additional taxonomy separating *hidden action* games from *hidden information* games. In the former, the agent undertakes actions unobservable to the principal, while in the latter the agent's actions may be observable but the reasons underlying his/her behaviour are not.

9.1.1 Moral hazard: the agency game

The principal–agent relationship is typically associated with the internal organization of firms, and in particular with the strategic interplay between stockholders and managers. That is, agency is seen as inherently linked to the separation between ownership and control of a firm. Large firms operate this separation, as shareholders (or simply owners) may not be trained to run a firm, and therefore find it appropriate to hire someone who possesses the required professional skills. Yet, this gives rise to a moral hazard issue, as the managers (i) possess information that owners cannot freely access (and if they did, they might not be able to interpret correctly), and (ii) shape their behaviour in accordance with the aforementioned information and their own objectives. The latter, unfortunately for shareholders, will not be aligned with profit maximization – in general – because managers are not holding significant shares of the firms' capital stock. Accordingly, managers might desire to maximize market shares, sales, etc., or simply shirk.

If approached from this angle, agency theory belongs to the theory of the firm and it can be considered basically as a formal approach to the problem of *opportunistic behaviour* outlined by Williamson (1975) and others. However, it is only fair to say that moral hazard and opportunism do affect any organization where the alignment of incentives along the hierarchical structure cannot be expected to emerge so easily.

A relatively simple model of an agency relationship can be outlined as follows.[1] Players are a principal, p, and an agent, a. For simplicity, I will set out illustrating the *hidden action* case.

The activity carried out by the agent produces a return $R(e)$, which increases with the agent's effort e. The latter is unobservable by the principal, although the latter knows how hard the agent should work to yield a certain return. This amounts to saying that the principal knows the functional form of $R(e)$ and therefore knows that, say, an effort $\tilde{e}$ is needed to generate a return $\tilde{R}$. The hiring contract specifies the agent's remuneration (or wage) $w(R)$ as a function of the principal's revenue R, which is the only thing that the principal can observe. The principal's objective function is $V(R(e) - w(R))$. If this is a firm and $R(e) - w(R)$ is profit, this means that the principal is happier the higher is the firm's profit. The agent wants to maximize the objective function $U(w, e)$, which is increasing with w and decreasing with e: that is, the agent dislikes effort. The strategic variables are w (controlled by the principal) and e (controlled by the agent).

The game unravels as follows:

1 The principal offers a contract specifying the wage level w to the agent.
2 The agent may or may not accept the job. He/she does so provided that the resulting utility $U(w, e)$ is at least as large as the so-called reservation utility $\overline{U}$ accessible to the agent on the market (i.e., outside this specific firm). If the agent signs the contract, he/she will produce an effort e unobservable by the principal.
3 This will generate a return $R(e)$ to the principal.

The resulting payoffs will be:

(a) $\pi_a = \overline{U}$ and $\pi_p = 0$ if the agent does not accept the contract; and
(b) $\pi_a = U(w, e) \geq \overline{U}$ and $\pi_p = V(R(e) - w(R))$ if the agent accepts the contract.

Now let's deal with the moral hazard problem implied by the unobservability of the effort. If the agent receives a fixed amount $\overline{w}$, he/she will sign the contract and then shirk. Therefore, the principal faces a binary choice between (i) offering a salary equal to zero (a trivial case in which the agent will never accept the job) and (ii) offering a salary that is indeed proportional to the revenues, $w(R)$. The solution of the principal's problem is a *forcing contract* whereby the agent's wage satisfies the conditions

$$U(w(R^*), e^*) = \overline{U} \quad \text{for all } R \geq R^*$$
$$U(w(R), e) < \overline{U} \quad \text{for all } R < R^* \tag{9.1}$$

That is, the principal sets a desired level of revenues R^* and writes a contract establishing that the agent will get a salary ensuring the attainment of the reservation utility $\overline{U}$ provided that the principal receives R^*. Otherwise, the agent is being punished, as the remuneration associated with any revenue level $R < R^*$ is insufficient to reach the agent's reservation utility.

Let's now introduce *hidden information* into the picture. In this case, revenues are determined by the agent's effort and a new element, usually labelled as the *state of the world*. The latter is a stochastic (or random) parameter ε, affecting also the agent's behaviour, so that $e(\varepsilon, w)$ and $R(e(\varepsilon, w), \varepsilon)$. In the existing literature, this aspect of the model usually involves the presence of a third player, *nature*, extracting the state of the world and showing it to one player only. The label 'state of the world' captures anything that can affect the performance of the agency we are looking at: for instance, in the case of a manufacturing firm, it can be an oil shock or a financial crisis. Both players know the probability distribution of the state of the world (i.e., how likely the occurrence of an oil shock tomorrow morning can be), but the actual value of ε is observable by the agent only. This private information intervenes in defining the intensity of the agent's effort e and complicates the task of the principal, who now has to write a contract specifying w under asymmetric information about e and ε at the same time. The above forcing contract clearly doesn't work any more: any unsatisfactory performance $R < R^*$ could be imputed by the agent to an unfavourable state of the world (or, bad luck), with the principal being altogether unable to understand whether the agent is telling the truth or just lying. Here, the effort is non-contractible not simply because it is unobservable but because of the hidden information about the state of the world, whereby the principal could never prove in a court of law that the agent has been shirking.

Given the uncertainty associated with the state of the world, we have to make an appropriate hypothesis concerning the attitude of both players towards risk. To begin with, suppose both agents are risk-neutral.[2] If so, then there exists an efficient solution that consists in *selling the pie*: the principal sells the firm at a given price P to the agent, who becomes residual claimant over the firm's profit flow generated by the agent's effort. The moral hazard problem has thus disappeared altogether because there is no longer any agency relationship: it is in the agent's best interest to choose an effort that maximizes the difference between the revenues $R(e(\varepsilon), \varepsilon)$ and the transfer price P, for any state of the world.

Assume instead (more realistically) that the principal is risk-neutral while the agent is risk-averse, so that the latter will request a higher remuneration under hidden information as compared to the case of hidden action. This generates a trade-off between the insurance against bad states of the world requested by the risk-averse agent and the need for the principal to design the most efficient incentive scheme for the agent in order to align the effort of the latter to the firm's objective.

The presence of an uncertain state of the world involves defining the objective functions in expected-value terms (as we did in Chapter 3 when we dealt with the Nash equilibrium in mixed strategies). Let the two players' expected utilities be

$$EV(R(e(\varepsilon, w), \varepsilon) - w(R(e(\varepsilon, w), \varepsilon))) \tag{9.2}$$

and

$$EU(R(e(\varepsilon, w), \varepsilon), w(R(e(\varepsilon, w), \varepsilon))) \tag{9.3}$$

respectively, for the principal and the agent.

The contract (or, the wage) must solve the principal's problem defined as the following constrained maximization programme. The principal chooses the wage w so as to maximize

$$EV(R(\hat{e}(\varepsilon, w), \varepsilon) - w(R(\hat{e}(\varepsilon, w), \varepsilon))) \tag{9.4}$$

subject to the constraints

$$EU(R(\hat{e}(\varepsilon, w), \varepsilon), w(R(\hat{e}(\varepsilon, w), \varepsilon))) \geq EU(R(e(\varepsilon, w), \varepsilon), w(R(e(\varepsilon, w), \varepsilon))) \tag{9.5}$$

for all $e \neq e^*$, and

$$EU(R(\hat{e}(\varepsilon, w), \varepsilon), w(R(\hat{e}(\varepsilon, w), \varepsilon))) \geq \overline{U} \tag{9.6}$$

The agent chooses the effort $\hat{e}$ so as to maximize his/her expected utility $EU(R(e(\varepsilon, w), \varepsilon), w(R(e(\varepsilon, w), \varepsilon)))$, instead of the principal's profits. This explains the presence of condition (9.5), the so-called *incentive compatibility constraint*. Condition (9.6) spells out the *participation constraint*: by accepting the contract, the agent must get at least the reservation utility $\overline{U}$.

I will not delve into the details of the solution of the principal's problem, as the mathematics involved goes well beyond the scope of the present book. From a purely qualitative point of view, I will confine myself to stressing that the solution of the constrained problem (9.4)–(9.6) yields a *second-best* equilibrium: by this, it is meant that the resulting wage $\hat{w}$ and effort $\hat{e}$, in general, will fall short of maximizing the principal's objective $EV(\cdot)$. A numerical example will help to grasp the intuition governing the rather intricate game going on between principal and agent.

9.1.2 Example

Consider a firm, with a principal and an agent. The labour input of the agent yields an output y, sold on the market at a fixed price $\bar{p} = 10$. The agent's effort can be either high, $\bar{e} = 2$, or low, $\underline{e} = 1$. The state of the world can be either good or bad, so that:

- if $e = \bar{e}$, we observe $\bar{y} = 100$ with probability $\mathfrak{p} = 2/3$ and $\underline{y} = 50$ with probability $1 - \mathfrak{p} = 1/3$; and
- if $e = \underline{e}$, we observe $\bar{y} = 100$ with probability $\mathfrak{p} = 1/3$ and $\underline{y} = 50$ with probability $1 - \mathfrak{p} = 2/3$.

That is, the occurrence of the unfavourable state of the world makes the attainment of the low output more likely, just by flipping the two probabilities over.

The agent is paid a certain wage w, so that the firm's (or equivalently, the principal's) profits are $\pi_p = \bar{p}y - w$, and for simplicity we may put $V = \pi_p$. Correspondingly, the objective function of the agent is defined as $U = \sqrt{w} - e^2$, showing a taste for salary and an aversion to effort. To complete the picture, let's fix the agent's reservation utility at $\overline{U} = 4$.

To begin with, we may quickly deal with the benchmark case in which information is symmetric, i.e., the agent's effort is indeed observable by the principal. If so, the latter can force the agent always to accept a fixed wage contract bringing the agent's utility down to the reservation level $\overline{U}$ for both $e = \bar{e}$ and $e = \underline{e}$:

- If $e = \underline{e} = 1$, then imposing $U = \overline{U}$ yields $\sqrt{w} - 1 = 4$ and therefore $w(\underline{e}) = 25$. As a result, the principal's payoff amounts to

$$V(\underline{e}) = \pi_p(\underline{e}) = \frac{2(500 - 25)}{3} + \frac{1000 - 25}{3} = \frac{1925}{3} \simeq 641.\bar{6} \tag{9.7}$$

- If $e = \bar{e} = 2$, then imposing $U = \overline{U}$ yields $\sqrt{w} - 4 = 4$ and therefore $w(\bar{e}) = 64$. As a result, the principal's payoff amounts to

$$V(\bar{e}) = \pi_p(\bar{e}) = \frac{500 - 64}{3} + \frac{2(1000 - 64)}{3} = \frac{2308}{3} \simeq 769.\bar{3} \tag{9.8}$$

Now suppose information is asymmetric. From (9.4)–(9.6), we may write the principal's constrained optimization problem as follows. The owner of the firm has to set two wage levels, $w(\bar{y})$ and $w(\underline{y})$,[3] to maximize

$$EV(\bar{e}) = \frac{500 - w(\underline{y})}{3} + \frac{2(1000 - w(\bar{y}))}{3} \tag{9.9}$$

under the incentive compatibility constraint

$$EU(\bar{e}) \geq EU(\underline{e}), \quad \text{i.e.,} \quad \frac{2\sqrt{w(\bar{y})}}{3} + \frac{w(\underline{y})}{3} - 4 \geq \frac{2w(\underline{y})}{3} + \frac{\sqrt{w(\bar{y})}}{3} - 1 \tag{9.10}$$

and the participation constraint

$$EU(\bar{e}) \geq \overline{U}, \quad \text{i.e.,} \quad \frac{2\sqrt{w(\bar{y})}}{3} + \frac{w(\underline{y})}{3} - 4 \geq 4 \tag{9.11}$$

This programme defines the wage schedule that must be offered to the agent in order for the latter to yield a high effort level. To identify the two unknown variables, $w(\bar{y})$ and $w(\underline{y})$, one has to use the notion that the constraints (9.10) and (9.11) will be satisfied at the margin, that is, as strict equalities. Accordingly, from (9.10) we obtain

$$\sqrt{w(\bar{y})} = \sqrt{w(\underline{y})} + 9 \tag{9.12}$$

which can be plugged into (9.11) to obtain

$$18 + 3\sqrt{w(\underline{y})} = 24 \tag{9.13}$$

so that $w(\underline{y}) = 4$ and therefore, using (9.12), $w(\bar{y}) = 121$.

Now we can simplify expression (9.9) to measure the expected value of the principal's payoff under asymmetric information:

$$EV(\bar{e}) = \frac{500 - 4}{3} + \frac{2(1000 - 121)}{3} = \frac{2254}{3} \simeq 751.\bar{3} \tag{9.14}$$

To conclude the exercise, note that

$$V(\bar{e}) > EV(\bar{e}) > V(\underline{e})$$
$$w(\bar{y}) > w(\bar{e}), \quad w(\underline{y}) < w(\underline{e}) \tag{9.15}$$

Under asymmetric information, the wage schedule inducing the agent to choose the high effort level is such that: (i) the firm's expected profits are lower than the level they would attain, for the same effort level, without asymmetry, but still higher than they would be in the unfavourable case with symmetric information; and (ii) the agent must be punished (respectively, rewarded) when the output is low (respectively, high) as compared to what he/she would receive when producing the low (respectively, high) effort under symmetric information.

9.1.3   Adverse selection: the market for 'lemons'

As anticipated above, the phenomenon of adverse selection is usually associated with the existence of asymmetric information across agents about some relevant features (quality, durability, etc.) of the goods being traded: producers/sellers know more than (potential) purchasers.

This is particularly true for *experience goods*,[4] whose quality is not only unknown *a priori* but is not immediately verifiable by purchasers, who have to use them for a while in order to ascertain their real quality. This typically holds for durable goods such as clothes, cars, TV sets, washing machines, cameras, etc., but also applies to non-durables (food, for instance). In any such cases, informative and persuasive advertising are relevant instruments in the hands of firms to send convincing messages (although not necessarily reliable ones) to consumers. A solution to the adverse selection problem is warranty: warranty goods are a subclass of experience goods with unobservable quality at the time of purchase, which are sold accompanied by a full refund warranty if the customer is not satisfied with them (goods must be brought or sent back to sellers after a reasonable amount of time, say a couple of weeks or so). By attaching a full refund warranty to experience goods, sellers are taking the burden of the risk associated with asymmetric information onto themselves, putting their customers in a condition of complete neutrality.

As an introductory illustration of the adverse selection problem, here we look at the famous model given by Akerlof (1970), concerning the second-hand car market. The role played by the state of the world in the agency problem is played here by the *type* of the good (or equivalently, of the seller). The players are a seller, s, and a (potential) buyer, b. The seller owns a car that he/she would like to sell to someone interested in a second-hand automobile. The quality θ of the used car is either high (H) or low (L); it is known to the seller, of course, but is unobservable to the buyer, who has solely information about the probability that any used car be good or a 'lemon': let's say that the probability of H is $\mathfrak{p}(H)$, while the probability of L is $\mathfrak{p}(L) = 1 - \mathfrak{p}(H)$. For simplicity, we may set

$$\mathfrak{p}(H) = \mathfrak{p}(L) = \frac{1}{2} \tag{9.16}$$

which, by the way, is also a sort of standard conjecture should the potential buyer be completely unaware of the probability of running across a lemon. Note that, assuming each seller has a single car, when *nature* extracts a car of either type from the whole population of second-hand cars, we could say as well that it is extracting a seller of either type from the corresponding population of sellers.[5]

Now, the buyer's problem is to bid a price for the car without observing its quality precisely. Of course, under symmetric information the fair prices for the two types would be $p_H = H$ and $p_L = L$, but the buyer cannot tell the type on the spot. Therefore, the best b can do is to offer

$$p_{E(\theta)} = \frac{H + L}{2} \tag{9.17}$$

the price corresponding to the expected quality level, given the probabilities attached to each type. The problem is that

$$H > p_{E(\theta)} > L \tag{9.18}$$

and consequently

- any type-*L* seller would be happy to sell, but then the buyer shouldn't buy because the car is a lemon; and
- all high-quality cars disappear from the market because their owners do not want to undersell them.

Of course, at this point lemons can be traded at their fair price. The repetition of the game (at least twice, but not necessarily more) may allow players also to trade good cars at the fair price, provided sellers know the identity of all sellers who left the market corresponding to $p_{E(\theta)}$. If so, then repetition yields a *separating equilibrium*, in which buyers can tell lemons from good cars, i.e., can separate the two types. I will come back to this notion when I illustrate incomplete information games.

The foregoing discussion can be summed up in the following remark.

REMARK 9.1   Under incomplete information, private (conflicting) incentives are bound to produce *market failures*. That is to say, the allocation of goods and services will be, in general, inefficient, as any uneven distribution of relevant information across the population of agents is bound to affect their incentives and the resulting price or wage schedule.

You may guess that a repeated game over an infinite horizon may open the way to a different equilibrium. Well, yes, this is indeed the case. The extension to allow for repeated interaction is illustrated below, in the section dedicated to incomplete information, where we will resort to a supergame in order to revisit the lemons problem. We will contrast this perspective with the simpler case where the game takes place only twice, to highlight the bearings of incomplete information and reputation on players' strategies and ultimately on the equilibrium outcome of the game.

9.2   Incomplete information

> Governments never learn.
> Only people learn.
>   Milton Friedman

Information incompleteness is usually captured in games by the assumption that some of the players do not know one or more relevant features characterizing the identity of other players. This is formalized by the notion of *player type*, as we have briefly seen in the game about the market for lemons. Uninformed players only know the exogenously given probability distribution of the rivals' types.

To solve games taking place under incomplete information, one has to construct a refinement of the Nash equilibrium concept, which was originally constructed for games of complete information. Thanks to a very smart transformation that was introduced by John Harsanyi (1967/68), an incomplete information game can in fact be solved as one in which information is complete but imperfect. The resulting solution concept is called Bayes–Nash or simply Bayesian equilibrium, and relies upon the so-called Bayes' rule. The rigorous derivation of the latter requires some basic concepts of statistical inference that I will confine to the appendix to this chapter. Here, it will be sufficient to explain the matter in purely intuitive terms.

Suppose you are facing the perspective of observing either of two mutually exclusive events, say, A or B, tomorrow. And you assign a prior probability to each of them. When the time comes, you indeed observe one, say, A, and not the other. Now, each of the two events may be driven by different causes, say, two, c_1 and c_2. In order to single out (if that is at all possible) the exact source of event A, you have to make a rational guess. What sort of guess? Your inference has to be based on the observation of the event. What you may do is just to assess the probability that the cause is indeed c_1 (or c_2) given your observation of event A. If the occurrence of A conveys some relevant information, then you may revise your priors in such a way that the occurrence of A reveals that, for instance, the cause is in fact c_1. In such a case, you obtain a *separating equilibrium* since you can tell c_1 from c_2. Otherwise, if A is not informative, your priors remain unchanged. In this case, you are 'trapped' in a *pooling equilibrium* because c_1 and c_2 remain, so to speak, overlapped, and you are unable to single out which one has caused event A.

A simple example will help. Suppose a boy has invited the girl he's secretly in love with to dinner at the best restaurant in town. He's asking himself whether she loves him back or not, and assigns to either event a probability of 1/2. (OK, this may sound overoptimistic, but after all it's just a trivial example.) Let's say, for simplicity, that she may reply either 'Yes, let me check my diary' or 'No, never in a lifetime'. The first answer, if objectively evaluated, is uninformative: dining does not really imply anything, does it? Therefore, priors remain 50/50 (or whatever they were). The second answer is informative, and our young man can revise his priors, which become zero and one, respectively.

If the basic idea is clear, we may pass on to re-examine the entry game that we met in Chapters 4 and 5, introducing a new element of incomplete information to make the model a bit more realistic. The game you are about to see dates back to Kreps and Wilson (1982a) and Milgrom and Roberts (1982a, b).

9.2.1 The entry game again

The structure of the constituent game is a variation on the theme proposed by Selten (1978) and Dixit (1979). Therefore, it replicates essentially the same structure as Matrix 5.5. Players are an incumbent, I, and an entrant, E. The entrant may either enter (e) or not (ne); the incumbent may either accommodate (a) or fight (f) entry. In the remainder, I will assume that the following payoff sequence holds for the entrant:

$$\pi_d > 0 > \pi_w^E \tag{9.19}$$

The interpretation is the same as in the game summarized by Matrix 5.5: if the incumbent accommodates entry, a duopoly with positive profits takes place; while if the incumbent triggers a price war, the negative payoff π_w^E denotes the entrant's loss.

Things are different for the incumbent, though. The latter can be one of two very different types: weak (W) or tough (T). If I is weak, then its payoffs in the various outcomes follow the ranking:

$$\pi_M > \pi_d > 0 > \pi_w^{WI} \tag{9.20}$$

with π_M and π_w^{WI} measuring monopoly profits and the losses caused by a price war, respectively. If instead the incumbent is of the tough type, the following payoff sequence

Matrix 9.1 Entry with incomplete information, I: I is weak

$$I$$

		a	f
E	e	$\pi_d; \pi_d$	$\pi_w^E; \pi_w^{WI}$
	ne	$0; \pi_M$	$0; \pi_M$

applies:

$$\pi_M > \pi_w^{TI} > \pi_d > 0 \qquad (9.21)$$

whereby in such a case the incumbent prefers a price war to the relaxed life of a duopoly equilibrium. This can be justified on the basis of the idea that an incumbent may bear significantly lower costs than a newcomer, and therefore can very much look forward to the perspective of starting off a price war in order to prey upon the entrant.

Matrix 9.1 portrays the constituent (or one-shot) game in the case of a weak incumbent, while Matrix 9.2 represents the game in which the entrant faces a tough incumbent.

As we already know, the original one-shot version of this game is fully symmetric, with $\pi_w^E = \pi_w^I = \pi_w$, and thus it has two pure-strategy Nash equilibria: (e, a), where a duopoly emerges at equilibrium; and (ne, f), with the incumbent retaining its monopoly power. The issue of multiplicity can be quickly solved by observing that, given the chain of inequalities in (9.20), the adoption of an aggressive behaviour is not a credible strategy because f is weakly dominated for the incumbent. Outcome (e, a) can instead be reached at the intersection of weakly dominant strategies by iteration and is also subgame perfect in the sequential play version of the game, with the entrant moving at the root of the corresponding Kuhn tree.

All of this would still apply here in the first version of the game (Matrix 9.1), if information were complete for the entrant. However, the second version (Matrix 9.2) gives rise to a completely different story. Here, the incumbent has f as the weakly dominant strategy, and therefore, when I is tough, (i) (e, a) is not an equilibrium (because $\pi_w^{TI} > \pi_d$), and (ii) (ne, f) is the *unique equilibrium*, because a tough entrant would always fight!

The entrant has an *a priori* distribution of probabilities over the incumbent's types, $\mathfrak{p}(T)$ and $\mathfrak{p}(W)$, with $\mathfrak{p}(T) + \mathfrak{p}(W) = 1$. Accordingly, from the standpoint of E, the one-shot game looks as in Matrix 9.3.

Matrix 9.2 Entry with incomplete information, II: I is tough

$$I$$

		a	f
E	e	$\pi_d; \pi_d$	$\pi_w^E; \pi_w^{TI}$
	ne	$0; \pi_M$	$0; \pi_M$

Matrix 9.3 The entrant's view of the game
under incomplete information

$$I$$

		a	f
E	e	$\pi_d; \pi_d$	$\pi_w^E; \mathfrak{p}(T)\pi_w^{TI} + \mathfrak{p}(W)\pi_w^{WI}$
	ne	$0; \pi_M$	$0; \pi_M$

When trying to figure out the outcome of the one-shot game, the entrant has to evaluate the behaviour of the incumbent. If E does so on the basis of the above matrix, the essential feature allowing the entrant to forecast I's strategy is the comparison between π_d and $\mathfrak{p}(T)\pi_w^{TI} + \mathfrak{p}(W)\pi_w^{WI}$. Two alternatives are admissible:

- If the prior probability of encountering a tough incumbent is low enough,

$$\mathfrak{p}(T) < \frac{\pi_d - \mathfrak{p}(W)\pi_w^{WI}}{\pi_w^{TI}} \qquad (9.22)$$

then

$$\pi_d > \mathfrak{p}(T)\pi_w^{TI} + \mathfrak{p}(W)\pi_w^{WI} \qquad (9.23)$$

This would imply that, as in Matrix 9.1, the game yields two equilibria, (e, a) and (ne, f), the former being in weakly dominant strategies.
- If, instead, the prior probability of encountering a tough incumbent is high enough to reverse both (9.22) and (9.23), the game has a unique equilibrium in weakly dominant strategies, (ne, f), as in Matrix 9.2.

The puzzling aspect of the story is the determination of priors and the problem of verifying whether these are indeed correct or not. To perform this twofold task, E must rely on historical information about the reaction of I to previous entries by other firms in the past (which may or may not exist or be readily available to E), and ultimately try his/her luck by entering the market and see what the incumbent actually does. So, the one-shot version of the game does not tell us that much, does it?

Accordingly, let's say that it is repeated over time, and, in order not to complicate the set-up too much, we may suppose that it takes place twice, with $t = 0, 1$. Both firms share the same time preferences, measured by the discount factor $\delta \in [0, 1]$. We are looking for a *subgame perfect Bayesian equilibrium*.

In order to characterize the equilibrium strategies, we have to comparatively assess the following discounted profit flows for the incumbent:

$$\pi_w^{KI} + \delta\pi_M \qquad (9.24)$$

with $K = T, W$ depending on the incumbent's type, and

$$\pi_d + \delta\pi_d = (1 + \delta)\pi_d \qquad (9.25)$$

Expression (9.24) is the flow of profits that I gets by fighting entry in the first period and remaining an uncontested monopolist in the second, while expression (9.25) is the alternative profit flow generated by accommodation in the first period, whereby we observe a duopoly in both.

The problem generated by incomplete information is therefore that E must: (i) formulate some prior probabilities attached to the two possible types of incumbent, $\mathfrak{p}(T)$ and $\mathfrak{p}(W)$; (ii) evaluate the expected profit flow that it can gain by entering at $t = 0$, in order to evaluate its own *a priori* incentive to enter; (iii) observe I's reaction to entry (i.e., whether the incumbent fights or not); and finally (iv) possibly use such observation – if informative about the type – to revise priors and tell exactly the incumbent's type.

Given the priors, the entrant's incentive is determined by the sign of

$$\mathfrak{p}(T)\pi_{\mathrm{w}}^{E} + \mathfrak{p}(W)\pi_{\mathrm{d}} = \mathfrak{p}(T)\pi_{\mathrm{w}}^{E} + [1 - \mathfrak{p}(T)]\pi_{\mathrm{d}} \tag{9.26}$$

If the above expression is positive, E will always enter, as expected profits are positive even in the presence of a tough incumbent. If instead (9.26) is negative, the entrant will be willing to enter only if the incumbent is weak and accommodates entry. But getting this piece of information is impossible if E stays out.

Of course, the incumbent is aware of the conjectures going on in the entrant's mind, and – under some specific condition that we are about to identify – can manipulate the game using the fog of information incompleteness affecting the entrant. The outcome of the repeated game crucially depends on two inequalities, capturing the incentives of the incumbent's two different types, respectively:

$$\pi_{\mathrm{w}}^{TI} + \delta\pi_{\mathrm{M}} > (1 + \delta)\pi_{\mathrm{d}} \tag{9.27}$$

and

$$\pi_{\mathrm{w}}^{WI} + \delta\pi_{\mathrm{M}} \overset{?}{\lessgtr} (1 + \delta)\pi_{\mathrm{d}} \tag{9.28}$$

Inequality (9.27), which is always verified due to the fact that $\pi_{\mathrm{w}}^{TI} > \pi_{\mathrm{d}}$ for type T, immediately implies the following remark.

REMARK 9.2   A tough incumbent will always fight entry.

Conversely, inequality (9.28) may take either sign, because, on the one hand, $\pi_{\mathrm{w}}^{WI} < \pi_{\mathrm{d}}$ on the basis of (9.20), but, on the other, $\pi_{\mathrm{M}} > \pi_{\mathrm{d}}$. Therefore, this implies the following remark.

REMARK 9.3   If $\pi_{\mathrm{w}}^{WI} + \delta\pi_{\mathrm{M}} > (1 + \delta)\pi_{\mathrm{d}}$, a weak incumbent will surely fight, mimicking the natural behaviour of type T. Conversely, if $\pi_{\mathrm{w}}^{WI} + \delta\pi_{\mathrm{M}} < (1 + \delta)\pi_{\mathrm{d}}$, a type-$W$ incumbent has no incentive to fight to mimic type T.

On these bases, one may draw the following conclusions, summarized in a theorem.

THEOREM 9.1   *If*

$$\mathfrak{p}(T)\pi_{\mathrm{w}}^{E} + [1 - \mathfrak{p}(T)]\pi_{\mathrm{d}} > 0 \quad and \quad \pi_{\mathrm{w}}^{WI} + \delta\pi_{\mathrm{M}} < (1 + \delta)\pi_{\mathrm{d}}$$

the weak incumbent doesn't fight (as opposed to a tough one, that always does). In such a case, we have a separating equilibrium, *since observing a (respectively, f) in the first period allows the entrant to deduce correctly that the incumbent is weak (respectively, tough). Accordingly, the second period takes place under complete information, as the entrant has been able to revise its priors to zero and one, depending on the incumbent's behaviour.*
If instead

$$\mathfrak{p}(T)\pi_{\mathrm{w}}^{E} + [1 - \mathfrak{p}(T)]\pi_{\mathrm{d}} < 0 \quad and \quad \pi_{\mathrm{w}}^{WI} + \delta\pi_{\mathrm{M}} > (1 + \delta)\pi_{\mathrm{d}}$$

a weak incumbent has an incentive to fight in the first period as if it were tough in order to retain monopoly power in the second period. As a consequence, in this case we have a

Matrix 9.4 Shall we fight for
natural resources?

		B	
		a	f
A	i	$R/2; R/2$	$\pi_w^A; \pi_w^B$
	ni	$0; R$	$0; R$

pooling equilibrium, *the entrant being unable to extract any information about the incumbent from its observed behaviour.*[6]

That is, by adopting an aggressive stance, the weak incumbent builds up a *reputation* of being tough – although it is not really the case – and the entrant is altogether unable to revise its prior probability conditional upon observed behaviour.[7]

9.2.2 *A wargame under incomplete information*

The fact that I have been dwelling upon the entry game for so long is most probably due to my being an economist. Do you want to know something about the role of incomplete information in a game belonging to the field of international relations? OK, here we go. Just take the above structure and tailor a new story around it.

The players are now two countries, *A* and *B*. At the outset, country *B* has monopoly rights over some valuable natural resource *R* (say, oil) located some place else, say, in a third country that is not playing a direct role in the game. Country *A* badly desires to hold a share of *R*, and is contemplating military aggression or an *invasion* of the third country (strategy *i*) to obtain it. Of course, *A* has the option not to do anything like that (strategy *ni*). Country *B* may either appease (*a*) or strike back (*f* for fight). The snapshot of the game is shown in Matrix 9.4.

Clearly, $R > R/2 > 0$. As in the previous game, the cornerstone of the model is one player's incomplete information about the opponent's type. Country *B* can be either of the tough type (in which case $\pi_w^B = \pi_w^{BT} > R/2$) or of the weak type (with $R/2 > 0 > \pi_w^B = \pi_w^{BW}$) and country *A* has priors $p(T)$ and $p(W)$, to be eventually revised upon observing the adoption of either strategy on the part of *B*. The solution follows exactly the same lines as in the previous game, and therefore I leave its details to you, as a useful exercise.

By the way, the same considerations apply in the extreme case, where the game is as in Matrix 9.5. Here, an invasion puts country *A* in full control of *R*, and leaves *B* with nothing. In this setting, we have either $R/2 > 0 > \pi_w^{BW}$ or $R > \pi_w^{BT} > 0$, and again we may observe either a separating or a pooling equilibrium depending on the relative size of country *B*'s discounted payoff flows. This formulation, although highly stylized, gives a hint about the mechanism that led to the First Gulf War, doesn't it?

Matrix 9.5 A 'radical' version of the same
game

		B	
		a	f
A	i	$R; 0$	$\pi_w^A; \pi_w^B$
	ni	$0; R$	$0; R$

9.2.3   *Product quality, reputation and signalling*

Let's go back to the 'lemons' market. In the field of industrial organization, a great deal of research has been carried out to understand firms' incentives to supply high-quality goods or conversely to distort quality downwards under asymmetric or incomplete information.

Leaving aside the possibility of introducing a warranty, a way out is offered by the repetition of the game (which, admittedly, is more viable for non-durables, like food, than for durables, like TV sets). The introduction of time may simplify the matter if quality is time-invariant, in which case consumers learn by experience all they need to know about the goods they purchase. Otherwise, if quality does change across periods, time considerably complicates the picture, as producers may want to modify quality from one period to another for strategic reasons, so that the adverse selection issue existing in the first period becomes a moral hazard problem from the second period onwards.

Let's consider first the simpler case in which quality is constant. One relevant question, among others, is whether sellers may adopt *introductory offers* – low prices – to convince consumers to buy. Can we expect a high-quality firm to make introductory offers? And, is it more or less keen on doing so than a low-quality firm?

A low price can attract more customers than a high one, and these customers may be happy to discover that the intrinsic value of their purchases is higher than expected at that low price. Yet, in order for price to convey information (and a *signal* of high quality), the gain generated by repeated trade must outweigh the cost advantage enjoyed by the low-quality firm. Under this condition, the signalling mechanism may indeed work in favour of the high-quality product.

A costly commitment usually associated with experience goods is advertising. In connection with it, the question is whether high-quality firms invest more or less money than low-quality firms in advertising campaigns. Casual observation suggests that high-quality goods need not be intensively advertised: have you seen many TV spots trying to convince anyone that Ferrari or Porsche are top-notch cars? I haven't. Yet, there is a lively discussion dating back to the early days of industrial organization theory about it, and also the empirical evidence is far from being conclusive – see Kihlstrom and Riordan (1984) and, for the empirical evidence, Caves and Greene (1996), *inter alia*.

The alternative version of the problem is that where quality can be modified over time. Any firm selling an experience good (in particular, if it's durable, and under the simplifying assumption that full or partial reimbursement is ruled out) has a strong incentive to decrease quality as time goes by, giving rise to a combination of adverse selection (in the first period) and moral hazard (thereafter). Repetition, however, can alleviate this phenomenon through the *reputation* that the seller may want to build up. In a repeated game, consumers may stop purchasing (for example, as a result of information obtained by word of mouth from previous purchasers) to punish a seller that is cheating on quality. A high profit margin thus emerges as a quality premium, in that (i) it rewards a reliable firm, that could otherwise be punished by consumers by simply not patronizing it any more, and (ii) it must be high enough to offset the temptation to decrease production costs by decreasing the quality level. These aspects are illustrated in two games we are about to see, introduced in the literature by Klein and Leffler (1981) and Shapiro (1983), respectively. The first is indeed a supergame over an infinite horizon, with *asymmetric information*.[8]

The players are a monopoly firm and a consumer. The firm supplies a single good of quality θ that, for simplicity, can be either high, H, or low, L. The unit cost of producing the high quality is c_H, while that associated with the low quality is c_L, with $c_H > c_L$. Time is $t = 0, 1, 2, 3, \ldots, \infty$, and $\delta \in [0, 1]$ is the constant discount factor. Assume that it takes a period to ascertain the actual quality of the good, so that the consumer gets to know at $t + 1$ the exact value θ of the unit purchased at t. The firm's reputation depends on the consumer's expectations about quality at time $t + 1$, which in turn are based on the quality observed at t.

To model the behaviour of players in the supergame, we may adapt the rules of Friedman's (1971) perfect folk theorem to the present set-up, in the following way. The firm starts off at $t = 0$ by supplying $\theta = H$, at a price equal to $p_H > c_H$. Subsequently, quality can be modified – i.e., brought down to L – and sold at the same price p_H, but only for a single period, as this spell is sufficient for the consumer to discover that he/she has paid a high price for a low-quality good, and therefore selling him/her quality L at p_H becomes impossible. More importantly, a single deviation by the seller is sufficient to destroy any reputation the firm may have, once and for all. If in any of the remaining periods the firm should 'claim' that the quality of its good is high again, the consumer would not be willing to buy anyway. This is, admittedly, a 'static' form of reputation, but is not too far from reality.

By now you should be familiar with the technicalities, so I'll go straight to the result. If the firm adopts the initial strategy forever, without any tricks, the resulting discounted flow of profits is

$$\Pi_H = (p_H - c_H)(1 + \delta + \delta^2 + \delta^3 + \cdots) = \frac{p_H - c_H}{1 - \delta} \tag{9.29}$$

We may compare it with the alternative payoff generated by a one-shot deviation towards L, sold at p_H. This yields the condition whereby the firm sticks to H forever and the consumer rewards its reliability:

$$\frac{p_H - c_H}{1 - \delta} \geq p_H - c_L \tag{9.30}$$

or, equivalently,

$$\delta \geq \frac{c_H - c_L}{p_H - c_L} \tag{9.31}$$

which surely belongs to the unit interval (i.e., it is an admissible value of the discount factor).

A reputational effect can also be singled out in a simple two-period framework. To see this, it suffices to introduce incomplete information by considering that a seller may be either *honest* or *dishonest*, and the consumer *a priori* can only formulate a probabilistic guess, and then possibly revise it on the basis of available information. The model is the same as above, with one additional detail only, concerning the description of consumer preferences, modelled as follows:

$$u = m\theta - p \tag{9.32}$$

The utility (or surplus) function u measures the difference between the satisfaction given by the consumption of a good of quality θ and the price, p. Parameter m is the so-called willingness to pay for quality, which is increasing with income: to put it bluntly, richer people are more keen on paying higher prices for superior quality levels than poor people.[9]

The rules of the game are as follows:

- at $t = 0$, the consumer 'trusts' the seller and buys one unit of the good, whose quality is fully revealed at the beginning of the second period; and
- at $t = 1$, if $\theta = L$ and $p = p_H$, the consumer concludes that the seller is dishonest, and doesn't buy (or recommends others not to); otherwise, if either $(\theta = L, p = p_L)$ or $(\theta = H, p = p_H)$, the consumer knows that the seller is honest, and buys again.

Accordingly, if the immediate cost advantage is higher than the discounted value of the maximum losses the seller will suffer because the consumer will not buy any more, cheating on product quality is a dominant strategy for the seller. Formally, this happens if the cost differential exceeds the profit margin, duly discounted:

$$c_H - c_L > \delta(\theta H - c_L) \tag{9.33}$$

whereby the seller acts dishonestly, the consumer becomes aware of that and a separating equilibrium obtains.

If, instead, the opposite holds, with $c_H - c_L < \delta(\theta - c_L)$, the reputational effect dominates the cost advantage, so that a dishonest type has an incentive to replicate the behaviour of the honest type, and will supply a high-quality good at p_H. As a consequence, observation is not informative and the consumer is unable to revise prior probabilities. This yields a pooling equilibrium.

9.3   Forward induction

> Successful investing is anticipating
> the anticipations of others.
>
> John Maynard Keynes

Thus far, throughout the volume, the *leitmotiv* underlying a player's rational choices has been backward induction. In any sequential play game, this criterion instructs a player to figure out the consequences of his/her decisions on the strategic behaviour of other players moving thereafter. So, what we do on this basis is to outline the unravelling of the game backwards, by endowing players with a forward-looking attitude.

There exists, however, an obvious alternative that, as we shall see in a short while, is very intriguing and can indeed provide startling new insights: it is the concept of *forward induction*, whereby each player – in order to forecast future developments, and ultimately the equilibrium outcome, starting from any node of a sequential play game – has to understand correctly the reasons why the game has reached that particular node and not others. To do this, every player has to interpret the behaviour of any other players that moved at previous nodes. This also entails that each player uses a more or less implicit *signalling mechanism* at every node, so as to transmit to others a rational hint about what he/she expects to happen in the remainder of the game.

The criterion of forward induction was introduced in a seminal paper by Kohlberg and Mertens (1986), and can be illustrated through a simple example. Observe the game tree in Figure 9.1.

Player 1, at the root, decides whether to stay out of the game (*o*), or enter it ('stay in', strategy *i*). If 1 stays out, the game is over and payoffs are 8 for player 1 and 20 for player 2. If 1

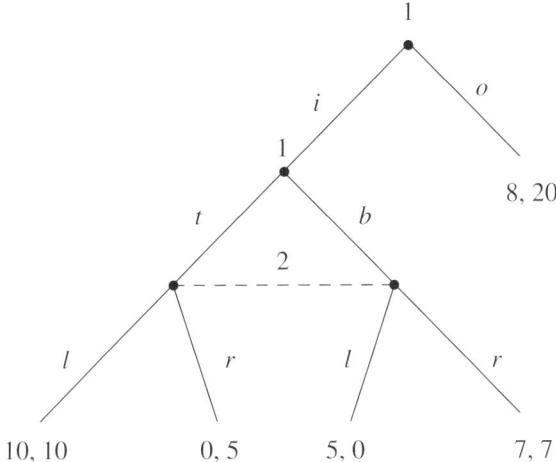

Figure 9.1 A forward induction game

chooses i, we have an imperfect information game starting at the node located at the second stage. Remember that, following Harsanyi, an incomplete information game can be solved as one of complete but imperfect information, so you may think of the tree as representing a game of incomplete information. The subgame starting at the second node from the top (after 1 has chosen i, and 2 has observed it) is a coordination game yielding two Nash equilibria, (t, l) and (b, r), that can be Pareto-ranked. Both players clearly prefer (t, l) to (b, r) but, under simultaneous moves, mistakes are indeed round the corner and one should account for the problematic solution in mixed strategies. Here is where forward induction helps to build up an equilibrium selection mechanism solving the coordination problem. If 1 is in, this fact transmits crucial information to player 2, as 1, by staying out, could get a payoff equal to 8 > 7. Therefore, player 2 should think: 'if 1 has chosen i at the root, it is because he/she will play t, as 10 > 8'. That is, forward induction permits players to exclude 'out-of-equilibrium beliefs': in this case, expecting 1 to choose b would be unreasonable because 1 can simply do better by playing o at the root.[10] Following this line of reasoning, forward induction selects (t, l). In other words, the behaviour of player 1 at the initial node ignites a signalling mechanism such that player 2 can figure out why 1 has played so and not in some alternative way.

Additionally, observe that the following remark may be made.

REMARK 9.4 Applying iterated deletion of (at least weakly) dominated strategies yields the same equilibrium as selected by forward induction, while in general the equilibria selected by backward and forward induction, respectively, do not coincide.

Proving the second part of Remark 9.4 is trivial. It suffices to observe that, since the intermediate nodes pertaining to player 2 in Figure 9.1 are not singletons, backward induction does not solve the problem posed by the multiplicity of Nash equilibria.

In order to sketch the proof of the first part of the remark, and appreciate its wide-ranging implications, you can just take a look at the strategic form of the same game, represented in Matrix 9.6, which has been built up by adding a third row to Matrix 3.13.

Matrix 9.6 Iterated dominance and forward induction

		2	
		l	*r*
	it	10; 10	0; 5
1	*ib*	5; 0	7; 7
	o	8; 20	8; 20

Strategy profiles *it* (*in*, then *top*) and *ib* (*in*, then *bottom*) describe the alternative choices of player 1 after deciding to stay in the game and interact with 2. Examine the matrix from the standpoint of 1: *ib* is strictly dominated by *o*, so the second row can be deleted. This is known to 2, that, on the remains of the matrix, finds out that *l* weakly dominates *r*. The last step consists in observing that the reduced form of the matrix (the 2×1 column where 2 surely plays *l* and 1 chooses between *it* and *o*) is such that 1 has in *it* a dominant strategy. Therefore, iterated dominance leads indeed to the same equilibrium outcome as forward induction.

The above structure can be applied to several different problems. One obvious candidate could be some version of the entry game we have seen many times – precisely for this reason, I won't spell out this example in detail. In a nutshell, if entry can give rise to two different equilibria, one with low profits and the other with high profits for both firms, and the outsider has the option to stay out and get more money than a price war would yield, then entry has a signalling value, inducing the incumbent to believe that the entrant points at a 'peaceful' duopoly equilibrium. If you are not already bored to death by entry games, you may see Brandts and Holt (1992). Instead, I will briefly illustrate a version of this game that fits the issue of entering an arms race or not.

9.3.1 Forward induction in arms races

The players are two countries, 1 and 2. Country 2 already has a nuclear arsenal, while coun-
try 1 is about to decide whether or not to build one for itself and become a member of the 'nuclear club', so to speak (strategies *i* for *in* and *o* for *out*, respectively). If country 1 enters the club, the next issue for both countries alike is whether these weapons should be used (strategy *a* for *attacking*) or not (strategy *p* for remaining *peaceful*). Matrix 9.7 illustrates the strategic form of the game.

The 2×2 submatrix that obtains by disregarding the third row is a coordination game with two Nash equilibria in pure strategies, (ip, ip) and (ia, ia). However, the payoff accruing to country 1 by choosing *o* is such that the strategy profile *ia* is dominated.[11] Using the same mechanism as above, one gets to the conclusion that iterated deletion of dominated strategies selects the equilibrium (ip, ip), ensuring perpetual peace through reciprocal nuclear deterrence. Now replace strategies and payoffs, wherever appropriate, in a game tree replicating

Matrix 9.7 Joining the 'nuclear
club'

		2	
		p	*a*
	ip	100; 100	−1000; 50
1	*ia*	50; −1000	0; 0
	o	60; 200	60; 200

Figure 9.1, and solve it by forward induction under the assumption that, if country 1 goes in, strategic interaction takes place simultaneously. You can easily verify that the equilibrium outcome selected by forward induction is again (ip, ip).

Further reading

At a textbook level, you may read more about incomplete and asymmetric information in Fudenberg and Tirole (1991), Myerson (1991), Montet and Serra (2003) and Rasmusen (2006). On learning processes in games, see Fudenberg and Levine (1998). For a very clear overview of the agency model, see the first chapter of Tirole (1988). The extension of the agency model to the case of many agents is covered in Alchian and Demsetz (1972) and Holmström (1982). The issue of using costly investments to signal product quality is discussed in Schmalensee (1978) and in Milgrom and Roberts (1986). Famous applications to models of economic policy, insurance, credit and labour markets can be found in Akerlof (1970), Spence (1973), Rothschild and Stiglitz (1976), Stiglitz and Weiss (1981) and Persson and Tabellini (1990). The debate on forward induction is duly accounted for in a wonderful book by van Damme (1991); see also van Damme (1989). For more on Bayes' rule in incomplete information games, see Mertens and Zamir (1985). For signalling games, see Banks and Sobel (1987) and Cho and Kreps (1987).

9.4 Appendix: Bayes' rule

The essence of Bayes' result can be laid out as follows. Any random event ϱ (generated, for instance, experimentally – remember Schrödinger's cat) occurs with some probability $\mathfrak{p}(\varrho)$ belonging to the unit interval. If ϱ occurs with absolute certainty, then $\mathfrak{p}(\varrho) = 1$. Conversely, if it's impossible, $\mathfrak{p}(\varrho) = 0$. The objective interpretation of probability holds it that

$$\mathfrak{p}(\varrho) = \frac{\text{number of favourable cases}}{\text{number of all possible cases}} \tag{9.34}$$

Now suppose there are n mutually exclusive events $\varrho_1, \varrho_2, \ldots, \varrho_n$. Their respective probabilities must sum up to one:

$$\mathfrak{p}(\varrho_1) + \mathfrak{p}(\varrho_2) + \mathfrak{p}(\varrho_3) + \cdots + \mathfrak{p}(\varrho_n) = 1 \tag{9.35}$$

Thus far, this is nothing really new. The additional step we have to take consists in grasping the definition and the implications of *conditional probability*. Imagine that event ϱ_1 may influence the probability of observing event ϱ_2. We shall indicate by $\mathfrak{p}(\varrho_2 \mid \varrho_1)$ the probability of observing ϱ_2 given the occurrence of ϱ_1, or simply the conditional probability of ϱ_2 given ϱ_1. Trivially, if

$$\mathfrak{p}(\varrho_2 \mid \varrho_1) = \mathfrak{p}(\varrho_2) \tag{9.36}$$

then ϱ_2 is independent of ϱ_1, and whether event ϱ_1 has taken place or not is altogether irrelevant for the probability of observing ϱ_2.

The *compound probability* that several independent events occur jointly is the product of their individual probabilities. We define by

$$\varrho_1 \cap \varrho_2 \cap \varrho_3 \cap \cdots \cap \varrho_{n-1} \cap \varrho_n \tag{9.37}$$

the intersection (that is what the symbol $\cap$ stands for) of the n events. The compound probability of observing these n events jointly is

$$\mathfrak{p}(\varrho_1 \cap \varrho_2 \cap \varrho_3 \cap \cdots \cap \varrho_{n-1} \cap \varrho_n) = \mathfrak{p}(\varrho_1) \cdot \mathfrak{p}(\varrho_2) \cdot \mathfrak{p}(\varrho_3) \cdots \mathfrak{p}(\varrho_{n-1}) \cdot \mathfrak{p}(\varrho_n) \qquad (9.38)$$

If these events are not reciprocally independent, we must resort to conditional probabilities. If you want to know the value of $\mathfrak{p}(\varrho_2 \,|\, \varrho_1)$, after a little proof that I will skip here, you will find out that

$$\mathfrak{p}(\varrho_2 \,|\, \varrho_1) = \frac{\mathfrak{p}(\varrho_2 \cap \varrho_1)}{\mathfrak{p}(\varrho_1)} \qquad (9.39)$$

When observing a generic event ϱ, which could be generated by two alternative causes h_1 and h_2, one may formulate the following crucial question: What is the probability that ϱ has been brought about by either h_1 or h_2, respectively? To find the right answer, we may rely on (9.39), and write

$$\mathfrak{p}(h_1 \,|\, \varrho) = \frac{\mathfrak{p}(h_1 \cap \varrho)}{\mathfrak{p}(\varrho)} \qquad (9.40)$$

Moreover, we have

$$\mathfrak{p}(\varrho) = \mathfrak{p}(h_1 \cap \varrho) + \mathfrak{p}(h_2 \cap \varrho) \qquad (9.41)$$

that is,

$$\mathfrak{p}(\varrho) = \mathfrak{p}(h_1)\mathfrak{p}(\varrho \,|\, h_1) + \mathfrak{p}(h_2)\mathfrak{p}(\varrho \,|\, h_2) \qquad (9.42)$$

We're almost there! Since

$$\mathfrak{p}(h_1 \cap \varrho) = \mathfrak{p}(h_1)\mathfrak{p}(\varrho \,|\, h_1) \qquad (9.43)$$

we have that, after plugging the relevant pieces into (9.40), this becomes

$$\mathfrak{p}(h_1 \,|\, \varrho) = \frac{\mathfrak{p}(h_1)\mathfrak{p}(\varrho \,|\, h_1)}{\mathfrak{p}(h_1)\mathfrak{p}(\varrho \,|\, h_1) + \mathfrak{p}(h_2)\mathfrak{p}(\varrho \,|\, h_2)} \qquad (9.44)$$

which captures Bayes' result (or, better, Bayes' theorem), whereby one can compute the *ex post* probability that the event ϱ be driven (or generated) by h_1.

In the jargon of the entry game formulated by Kreps and Wilson (1982a) and Milgrom and Roberts (1982a), we may set $h_1 = T$, $h_2 = W$ and $\varrho = f$.

The entrant revises the priors $\mathfrak{p}(T)$ and $\mathfrak{p}(W)$ guided by the incumbent's observed behaviour, so that the *ex post* conditional probability that the incumbent is indeed of the tough type, having observed aggressive behaviour, is given by the following expression:

$$\mathfrak{p}(T \,|\, f) = \frac{\mathfrak{p}(T)\mathfrak{p}(f \,|\, T)}{\mathfrak{p}(T)\mathfrak{p}(f \,|\, T) + \mathfrak{p}(W)\mathfrak{p}(f \,|\, W)} \qquad (9.45)$$

If the incumbent adopts strategy a, the entrant knows for sure that I is weak, and therefore sets $\mathfrak{p}(W \,|\, a) = 1$. This yields the separating equilibrium. If, conversely, the incumbent chooses f, priors cannot be revised and the pooling equilibrium obtains.

10 Bargaining and cooperation

We get it on most every night
when that moon is big and bright
it's a supernatural delight
everybody's dancing in the moonlight
we get everybody here is out of sight
they don't bark and they don't bite
they keep things loose they keep it tight
everybody's dancing in the moonlight

Sherman Kelly (King Harvest),
Dancing in the moonlight

The world we're in can hardly be described by a series of cooperative games. If you have gone through the previous chapters, you should be aware by now of how little cooperation one may hope to observe in economics and politics, as well as in other areas of life. Having said that, there are many aspects of economics, politics and international relations where negotiations are most relevant, and therefore a book on game theory is incomplete if it doesn't contain an illustration of cooperative and bargaining games. However, explaining and understanding these games is no easy task, because of the amount and nature of formal analysis that they require even at an introductory level. Here, to keep the exposition in line with the remainder of the book, I will lay out the basics of this very relevant part of game theory, and propose two simple examples that give a flavour of the range of potential applications.

Starting from the publication of the book by von Neumann and Morgenstern (1944), for about 30 years research efforts were very intense in this area. Indeed, when game theory was still relatively young, issues related to coalition formation and the design of stable agreements across agents that could instead pursue conflicting goals were hot topics on the agendas of most game theorists.[1]

There were two main routes along which research was being developed: the *positive* approach and the *axiomatic* one. While the former is directed to shed light on the relationship between cooperative games and the theory of general equilibrium in perfectly competitive economic systems, the latter is characterized by a normative nature and investigates the operational features of different solution concepts based upon the smallest possible set of unproven intuitive axioms.

The essence of cooperative as well as bargaining games is that agents are supposed to be able to design and implement strategies leading to a Pareto improvement as compared to the outcome of the purely non-cooperative version of the same game.

The problem then consists not in defining the basic layout of the game (players, their individual payoffs and their admissible strategies can be defined in the same way as in Chapter 2), but rather in understanding whether this cooperation is feasible and, above all, *stable*. Cooperation, in itself, means literally that individuals are assumed to choose their strategies in a cooperative way. Bargaining has to do with the way they negotiate the partition of gains across themselves – a peculiar activity that may be extremely conflicting.

A conceptual issue affecting the idea that agents may explicitly cooperate is the associated need of postulating a common interest, which in turn points in the direction of discarding altogether the alternative idea that each agent rationally seeks the maximization of his/her own satisfaction, one of the cornerstones of (non-cooperative) game theory. This need was first highlighted by Nash (1951). As a result, the development of this branch of bargaining game theory is known as the *Nash programme*. This prompted the investigation of cooperative games using the very same toolkit as employed for non-cooperative games, giving rise to a branch of the theory focusing on *strategic bargaining* games, whose highest peak is the work of Rubinstein (1982). Another way of putting it is that strategic bargaining games can be a conceptual bridge between non-cooperative games and purely cooperative ones. The difficulty with this approach to bargaining games is that it very often generates a multiplicity of equilibria (requiring then refinements, as for the non-cooperative games we have reviewed thus far). The axiomatic approach eliminates this problem by introducing sensible and desirable properties in the form of axioms, following Nash (1950b, 1953). Of course, in such a case the critical aspect is the nature and specification of these axioms. To avoid the expositional and technical difficulties associated with the strategic bargaining approach, here I will only tell you about the axiomatic approach to bargaining, and the Nash programme in which it nests.

10.1   Bargaining games: the axiomatic approach

> A peace is of the nature of a conquest;
> for then both parties nobly are subdued,
> and neither party loser.
>
> William Shakespeare,
> *Henry IV*

As anticipated above, the axiomatic approach is based upon a set of intuitive claims (the axioms) describing some essential features that must be met by the set-up under examination (in our case, the game) and then draws some predictions relying on such intuitive properties. More specifically, what Nash (1950b, 1953) does is use selected axioms to prove that they lead to a unique outcome satisfying all of them at the same time. This is the so-called Nash bargaining solution.

10.1.1   The Nash bargaining solution

To keep things as simple as possible, consider a two-player variable-sum game in which each agent obtains a payoff or utility measured by $u_i(s)$, $i = 1, 2$, for any given outcome $s = (s_1, s_2)$. The game is sketched in the utility or payoff space in Figure 10.1, which closely resembles Figure 5.2.

Let the pair $(u_1(d), u_2(d))$ identify the utility levels accruing to each agent if they disagree. This is the *disagreement* or *threat* point D, which players can always reach non-cooperatively if bargaining breaks down and they fail to reach an agreement – in fact, it can

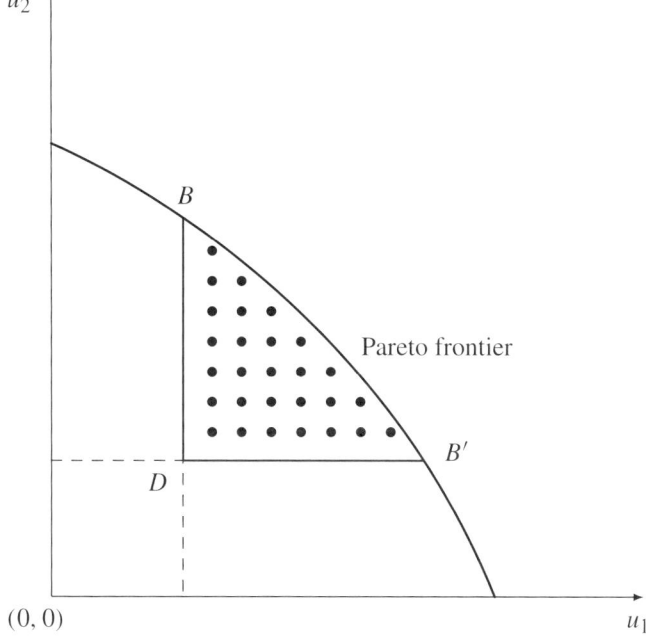

Figure 10.1 The feasible set

be thought of as the Nash equilibrium of the non-cooperative version of the same game. The dotted region DBB' defines the *feasible set*, i.e., the set of all outcomes that (i) are Pareto-efficient as compared to the disagreement point, and (ii) can be reached through the bargaining process starting from the same point.

No details about the bargaining process or the agreement between players are given. What Nash does is just to outline the equilibrium solution, which, in his view, must drive players onto the Pareto frontier of the feasible set. The *Nash bargaining solution* (*NBS*) is attained by maximizing the so-called *Nash product*:

$$(u_1 - u_1(d))(u_2 - u_2(d)) \tag{10.1}$$

with respect to u_1 and u_2. The individual utilities $(u_1(d), u_2(d))$ associated with the disagreement point being given, the Nash product (10.1) changes with u_1 and u_2 in such a way that it draws a map of curves that are convex towards the origin of the axes, as in Figure 10.2.

The maximization of (10.1) must be compatible with the constraint, namely, that players have to choose a point within the feasible set, or, at most, along its frontier. Therefore, (10.1) is indeed maximized when agents locate at the tangency point between the Pareto frontier of the feasible set and the highest possible curve generated by (10.1) that is compatible with the feasible set. This is point *NBS* in Figure 10.2. I will label its coordinates as (u_1^*, u_2^*).

The Nash bargaining solution relies on the following axioms.

- *Individual rationality*: $u_i^* \geq u_i(d)$ for $i = 1, 2$; trivially, this says that the Nash bargaining solution must yield at least the same utility as the disagreement point.

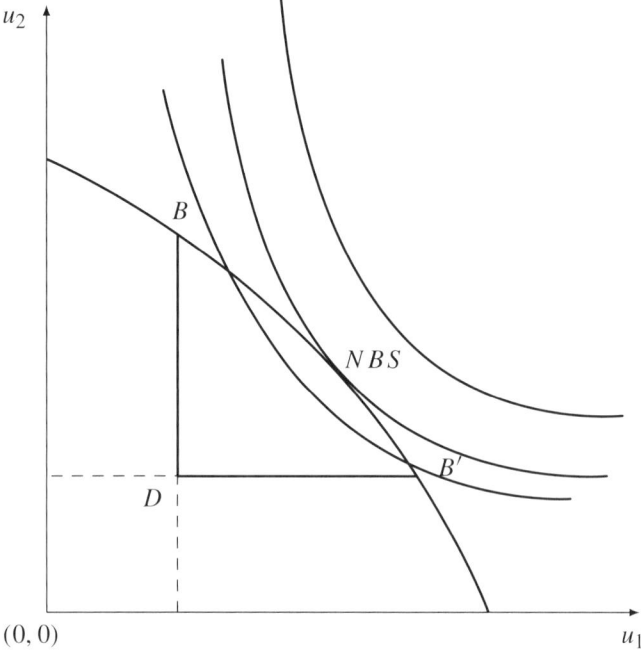

Figure 10.2 The Nash bargaining solution

- *Invariance*: the Nash bargaining solution is independent of the scale used to measure the players' utility; that is, changing from metres to yards or from kilograms to pounds does not affect the solution.
- *Pareto optimality*: if $(u_1, u_2) > (u_1^*, u_2^*)$, then (u_1, u_2) cannot belong to the feasible set; this axiom says that neither player can unilaterally improve upon the Nash bargaining solution without damaging the other.
- *Independence of irrelevant alternatives*: the Nash bargaining solution is unaffected by excluding any feasible pairs of utilities not selected by players.
- *Symmetry* or *anonymity*: identifying players is irrelevant to the solution; this axiom could be also read as 'one man, one vote'.

Nash (1950b, 1953), in a theorem whose details I will skip, proves that his bargaining solution is the only one meeting the whole of the axioms listed above. It is worth noting that this axiomatic approach is common to Arrow's impossibility theorem (Arrow, 1951) that we encountered in Chapter 6. This is not happening by chance, of course: the axiomatic approach has a strong tradition in twentieth-century mathematics in general.

   To appreciate how the Nash bargaining solution can be used in practice, one may resort to a simpler case where the efficiency frontier is a straight line. We already know a case that has this feature: the cartel we have investigated in Chapter 5. There, we assumed that firms were able to stabilize collusion along the supergame; here, we may revisit the same problem assuming that firms resort to bargaining, no matter how long is the time horizon (in fact, a one-shot game will do).

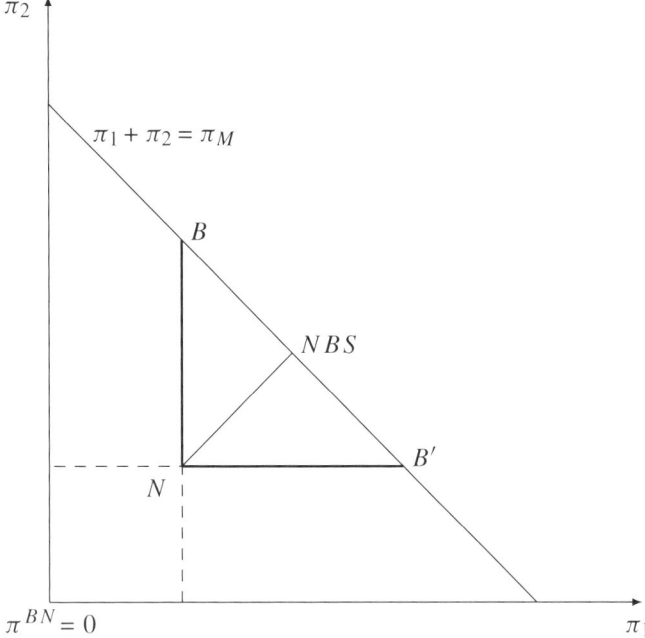

Figure 10.3 Example: the linear frontier case in duopoly

If firms are identical, the disagreement point has coordinates (π^N, π^N), where utilities coincide with profit levels, and the latter are measured by the non-cooperative Nash equilibrium profits. This is a point like N in Figure 10.3, where the origin identifies the Bertrand–Nash equilibrium with zero profits.

The Nash bargaining solution can be calculated by splitting equally the additional profit pie generated by the cartel, using the following simple rule (see Muthoo, 1999):

$$\pi^* = \pi^N + \frac{\pi_M - 2\pi^N}{2} = \frac{\pi_M}{2} \tag{10.2}$$

which of course yields 50 per cent of monopoly profits to each firm. More interesting is the case where the disagreement point (π_1^N, π_2^N) is asymmetric, with $\pi_1^N \neq \pi_2^N$ (due, for instance, to asymmetric technologies or some degree of product differentiation). Also, in this case, we can resort to the analogue of (10.2) to calculate the profits accruing to each firm in the cartel:

$$\pi_i^* = \pi_i^N + \frac{\pi_M - \pi_i^N - \pi_j^N}{2} \tag{10.3}$$

although of course this results in an asymmetric partition of monopoly profits reflecting the asymmetry in the disagreement point. Observe that collateral or *side payments* should be allowed here as well as in other similar cases where some degree of asymmetry operates. The technical label used to express this idea is that these must be games with *transferable utility*, where players can use some form of common currency (like money) to transfer payoffs between each other.

The Nash bargaining solution has generated a lively debate and a large amount of research. Possibly the most startling product of this research is the so-called *Kalai–Smorodinsky solution* (Kalai and Smorodinsky, 1975), which adopts an alternative to the axiom of independence of irrelevant alternatives, called the *individual monotonicity axiom*. The latter requires that, if the feasible set is expanded, and this expansion benefits specifically one of the players, then also the associated bargaining solution must benefit the same player. Subsequently, Kalai (1977) introduced the *egalitarian solution*, which drops the invariance axiom and can be considered as a compromise between the Nash bargaining solution and the Kalai–Smorodinsky solution, since it imposes that

$$u_1^* - u_1(d) = u_2^* - u_2(d) \tag{10.4}$$

i.e., bargaining must result in equal gains for all players as compared to the disagreement point.

Furthermore, this stream of research has shown that, under some particular but plausible conditions, which are likely to be verified often, the Nash bargaining solution and its non-cooperative counterpart due to Rubinstein (1982) indeed coincide (see Binmore *et al.*, 1986, 1992, *inter alia*).

These bargaining solutions can be used in games with many players (see, e.g., Harsanyi, 1977). In such cases, however, subgroups of players can form coalitions to safeguard the interests of their members. This is the topic treated in the next section.

10.2   Cooperative games: a matter of coalitions

> All the people like us are We,
> and everyone else is They.
>
>   Rudyard Kipling

The bargaining approach outlined above is based on the idea that *individuals* activate bargaining processes. The main body of cooperative game theory instead adopts the view that individuals may join to form coalitions, the latter becoming the effective actors in a cooperative game.

10.2.1   *The positive approach to cooperation: the core*

This approach postulates that players will adopt strategies apt to yield Pareto-efficient outcomes. To this end, they can build up coalitions through binding commitments that, in view of the design of the cooperative game, are credible. When examined from this angle, the main task that cooperative game theory has to perform is to identify the appropriate mechanism by which additional gains generated by cooperation are to be distributed over the population of players. The main solution proposed by game theory in this respect is the *core* (Gillies, 1959).

To fix ideas, think of a simplified economic system where there is only trade but no production. This is usually called a *pure exchange economy*, where, corresponding to a certain list of relative prices, a number of agents (or coalitions thereof) trade all or part of their respective endowments of goods or assets. This toy economy, for all those who are familiar with microeconomics, is the basic framework usually presented to first-year students to illustrate for them for the first time the general equilibrium model dating back to Edgeworth (1881).

By the very fact of entering this trading game – alone or as a member of a coalition – every agent will get a payoff. To begin with, let $v(i)$ measure player i's individual *worth* or *value* when the latter acts on his/her own. This, bluntly speaking, refers to the case of a trivial 'one-man-band' coalition. Any non-trivial coalition must satisfy two properties. Intuitively, there must be an incentive for each individual to join a coalition and obtain something more than he/she would get if standing alone. Moreover, any non-trivial coalition must be *superaddi-tive*, i.e., the coalition resulting from the merger of two previously independent coalitions must be able to attain more than the sum of payoffs or utilities attainable by its constituent parts as autonomous coalitions: cooperation always helps.

Then, define the *grand coalition* as $\mathfrak{C}$, i.e., the coalition gathering all the n players involved in the game. The value of the grand coalition is

$$v(\mathfrak{C}) = \max_{s_{\mathfrak{C}}} \sum_{i=1}^{n} u_i(s_{\mathfrak{C}}) \tag{10.5}$$

where $s_{\mathfrak{C}} = (s_1, s_2, s_3, \ldots, s_n)$ is the list of strategies of all players in the grand coalition and $u_i(\cdot)$ is the utility or objective function of player i. The value $v(\mathfrak{C})$ is also called the characteristic function of coalition $\mathfrak{C}$ and measures the total amount of transferable utility that can be distributed across its members via some agreeable rule. Any sub-coalition (sometimes called *proper* coalition) of size $\mathfrak{c}$, with $1 < \mathfrak{c} < \mathfrak{C}$, also has a characteristic function,

$$v(\mathfrak{c}) = \max_{s_{\mathfrak{c}}} \sum_{j=1}^{c} u_j(s_{\mathfrak{c}}) \tag{10.6}$$

which can be thought of as the maximum payoff attainable by its members irrespective of the behaviour undertaken by outsiders $\mathfrak{c}+1, \mathfrak{c}+2, \ldots, n$. Equivalently, it can also be understood as measuring the total payoff that coalition $\mathfrak{c}$ can guarantee (and distribute to) its members if the outsiders play against this coalition, i.e., behaving so as to minimize the coalition's worth.

At this point, we have to introduce the concept of *imputation*: given v and $\mathfrak{C}$, the imputation $\mathfrak{J}(v, \mathfrak{C})$ is a list of payoffs

$$\pi_{\mathfrak{C}} = (\pi_1, \pi_2, \ldots, \pi_n) \tag{10.7}$$

such that

$$\pi_i \geq v(i) \quad \text{for all } i \tag{10.8}$$

and

$$\sum_{i=1}^{n} \pi_i = v(\mathfrak{C}) \tag{10.9}$$

Condition (10.8) establishes that the imputation must be individually rational, i.e., each agent must have an incentive to join the coalition. Condition (10.9) spells out a collective rationality requirement. The latter also embodies Pareto efficiency, as players cannot achieve more than what is allowed by the grand coalition.[2]

Now pay attention to the following crucial definition.

DEFINITION 10.1 The core $C(v, \mathfrak{C})$ of the game, if it exists, is the subset of the set of imputations $\mathfrak{I}(v, \mathfrak{C})$ that is non-dominated by any blocking coalition $\mathfrak{c}_b$.

The meaning of this definition may indeed look a bit obscure at first sight. It can be spelled out more clearly in these terms. The core must satisfy both individual and group rationality, i.e., there must be an incentive for each agent to join the grand coalition, and there must be no proper sub-coalition in a position to outperform the grand coalition, i.e., able to grant its members at least as much as they could get by adhering to the grand coalition. If such a coalition exists, it can in fact get along on its own and *block* the grand one. Consequently, you may understand the core as the subset of all possible imputations that are Pareto-efficient and cannot be improved upon by any (proper) sub-coalition.

Note that, unfortunately, the core may not exist – that is, there are games in which the set of Pareto-efficient and non-dominated imputations may in fact be empty. Without discussing the properties that a game must possess to ensure the presence of a non-empty core and delving into the details of the formal proof,[3] I will confine myself to saying that, in general, the core will be non-empty in games where any imputations $v(\mathfrak{c})$ generated by proper coalitions do not dominate the improper 'one-man-band' coalition, $v(\mathrm{i})$.

A relevant issue connected with the emptiness of the core is whether this fact implies that cooperation is altogether impossible and, in a sense, invalidates the idea that a group of individuals could be guided by some plausible form of collective rationality. If a society or community cannot collectively improve upon the outcome yielded by conflicting individual interests, then atomistic behaviour is inherently costly and should be replaced by planning: a centralized decision-making process controlled by a single benevolent dictator. Of course, the one in control may turn out not to be benevolent, and Orwell's *Big Brother* is there to remind us about the unpleasant implications of this perspective.

10.2.2   The axiomatic approach to cooperation: the Shapley value

The basic idea behind the Shapley value (Shapley, 1953) is to define a numerical index measuring the value that each single player assigns to the fact of taking part in a game, individually or as a member of a coalition. In other words, the Shapley value is a list (a vector, to be precise) of numbers, each one measuring the actual coalitional power of a specific player.

Let $N = 1, 2, 3, \ldots, n$ be the set of all players involved, and $\mathfrak{c}$ be a proper coalition of size $\mathfrak{c}$ including player i, with a characteristic function $v(\mathfrak{c})$, while $v(\mathfrak{c} - i)$ measures the value that the same coalition would attain by excluding player i. Accordingly, the marginal (individual) contribution of i to this coalition is

$$\Delta_i = v(\mathfrak{c}) - v(\mathfrak{c} - i) \tag{10.10}$$

The Shapley value defines a criterion[4] to be used to impute the reward or payoff to each player on the basis of his/her individual contribution Δ_i to the coalition.

Shapley's construction proceeds by axiomatization. The value $V_i(\mathfrak{c})$ for player i has to respect the following axioms.

- *Group rationality*: $\sum_{i=1}^{c} V_i(c) = v(c)$, i.e., the allocation of payoffs across individuals exhausts what is collectively achieved by the coalition.
- *Equal treatment*: if any two players i and j in the coalition are 'equivalent' in terms of their individual contributions to the coalition, then $V_i(c) = V_j(c)$, i.e., they receive the same reward.
- *Null player condition*: if the marginal contribution of player i coincides with the value of the same player as a 'one-man-band' coalition, whereby $\Delta_i(c) = v(c) - v(c - i) = v(i)$, then such a player should be rewarded accordingly, with $V_i(c) = v(i)$.

Relying on this set of axioms,[5] the Shapley value is calculated taking into account that player i can adhere to several proper coalitions. Consider all the possible coalitions whose size is c. The Shapley value is

$$V_i(c) = \sum_{c \subseteq N} \frac{(c - 1)!\,(n - c)!}{n!} \Delta_i(c) \tag{10.11}$$

This is nothing other than a weighted average of the marginal contributions that i could bring to all the possible coalitions of size c that can be formed by players belonging to N.[6] It should be noted that the core and the Shapley value do not coincide in general, and are not even necessarily reciprocally compatible: if the core is non-empty, the Shapley value may or may not belong to the core. The reason is to be found in the instability caused by the incentive for secessions of existing coalitions.

10.3 Examples

It is now time to see some applications to real-world problems. The three games contained in the last part of this chapter deal with topics belonging to international relations and environmental economics, respectively.

10.3.1 *Mutual defence: alliances, the core and the Shapley value*

Building up, and eventually enlarging, alliances to share the defence burden is a topic of great interest in the field of international relations. The creation of NATO and its enlargement process is probably the most striking example of all.[7] The following game is based on Gardner (1995) and Sandler (1999).

Suppose that three contiguous countries, A, B and C, are contemplating the possibility of signing a pact to create a mutual defence alliance. In the benchmark case, each country is a unit square with four borders of unit length to protect from an external threat in any direction. The three countries are configured as in Figure 10.4 so as to form a rectangle.

For simplicity, the cost of defending a country or the alliance against an attack from abroad is taken to be proportional to the extent of the perimeter. In particular, protecting each side of these countries involves a unit cost. Consequently, $v(i) = -4$, $i = A, B, C$, is the cost that country i has to bear when defending its borders on its own. Similarly, $v(a)$ represents the cost of protecting an alliance of countries in a coalition of size a.

Now suppose that the presidents/premiers and foreign affairs ministers of these countries sit around the same table to play a cooperative game whose subject is the construction of an alliance possibly taking the form of the grand coalition $\mathfrak{A}$, whose value is

$$v(\mathfrak{A}) = v(A, B, C) = -8 \tag{10.12}$$

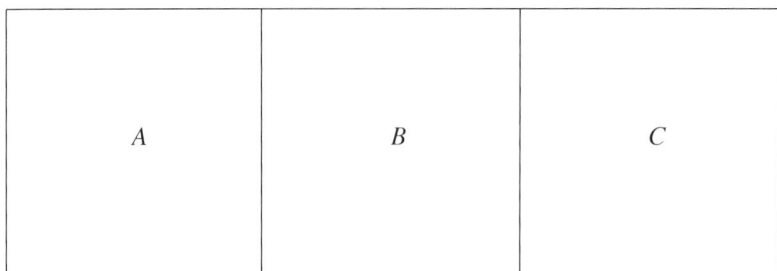

Figure 10.4 Sharing the defence burden: the benchmark case

Note that all three countries:

- are aware that three proper coalitions are possible, with values corresponding respectively to

$$v(A, B) = -6, \quad v(B, C) = -6, \quad v(A, C) = -8 \tag{10.13}$$

- know that (i) country B is in the best position to form any coalition because it shares two sides with the other countries, and (ii) the coalition involving countries A and C has no gains at all to offer as these countries are not contiguous with each other, and therefore we may expect the alliance between A and C to play no role at all in negotiations.

If any proposal on a *grand alliance* giving rise to a payoff allocation $u = (u_A, u_B, u_C)$ is on the table, in order to belong to the core, it must satisfy the condition

$$v(\mathfrak{a}) \leq u_i, \quad i = A, B, C \tag{10.14}$$

for any possible coalition $\mathfrak{a}$ (otherwise countries in $\mathfrak{a}$ would be better off in $\mathfrak{a}$ and reject the proposal). This translates into the following set of $2^3 - 1 = 8 - 1 = 7$ conditions:

$$
\begin{aligned}
v(i) &= -4 \leq u_i, \quad i = A, B, C \\
v(A, B) &= -6 \leq u_A + u_B \\
v(B, C) &= -6 \leq u_B + u_C \\
v(A, C) &= -8 \leq u_A + u_C \\
v(\mathfrak{A}) &= -8 \leq u_A + u_B + u_C
\end{aligned}
\tag{10.15}
$$

Now, using the transformation

$$u'_i = u_i + 4 \tag{10.16}$$

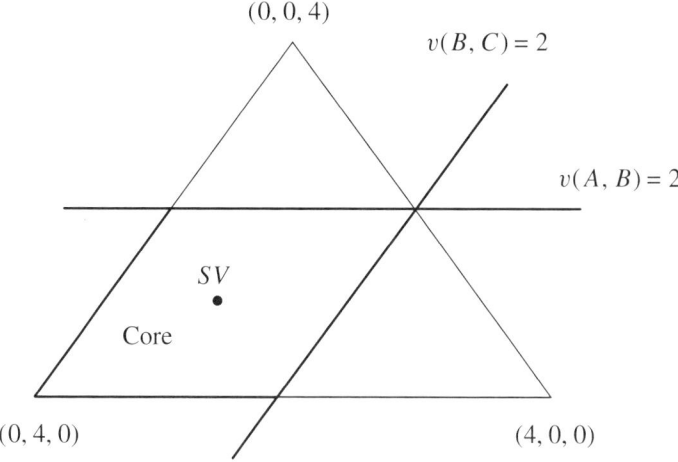

Figure 10.5 The core and Shapley value of the benchmark case

the list (10.15) can be rewritten as follows, so as to measure how much is actually being saved on total defence costs by any coalitions:

$$v(i) = 0 \leq u'_i, \quad i = A, B, C$$
$$v(A, B) = 2 \leq u'_A + u'_B$$
$$v(B, C) = 2 \leq u'_B + u'_C \quad\quad\quad (10.17)$$
$$v(A, C) = 0 \leq u'_A + u'_C$$
$$v(\mathfrak{A}) = 4 \leq u'_A + u'_B + u'_C$$

which can be easily used to investigate geometrically the solution of this cooperative game. This is done in Figure 10.5.

The triangle represents the Pareto-efficient plan along which players can locate themselves through bargaining. It is a surface because of the presence of three players. Point $(0, 4, 0)$ identifies the best agreement for country B; likewise, points $(4, 0, 0)$ and $(0, 0, 4)$ represents the best agreements for countries A and C, respectively.

Any allocation corresponding to a point of the triangle above the horizontal line $v(A, B) = 2 = u'_A + u'_B$ is dominated by the alliance involving A and B and excluding C. Similarly, any allocation corresponding to a point located to the south-east of the positively sloped line $v(B, C) = 2 = u'_B + u'_C$ is dominated by the proper coalition involving B and C only. The rhombus outlined by the intersections between these lines and the perimeter of the triangle is the core, whose points are undominated by either proper coalition.

Furthermore, applying (10.11), we may identify the Shapley value as the triple $(1, 2, 1)$, which exhausts the total payoff equal to 4 by imputing it entirely to the players, with the highest payoff accruing to country B reflecting the coalitional value of the latter. It is also worth noting that the point corresponding to the Shapley value does belong to the core: in particular, it is identified geometrically as the intersection of the diagonals of the rhombus at SV, in Figure 10.5.

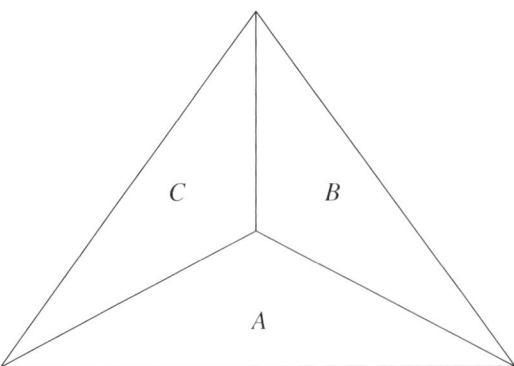

Figure 10.6 A triangular world

So far, I have briefly reiterated what you may find in Gardner (1995, pp. 401–5) and in Sandler (1999, pp. 731–4). An interesting alternative is that where all countries have two sides in common, which makes them entirely symmetric. This is the case illustrated in Figure 10.6, where this region of the world is triangular, each country being itself a triangle nested in the first.

Repeating the same exercise as in the benchmark case, one obtains (I will skip the payoff transformation procedure for brevity):

$$v(i) = 0 \leq u'_i, \quad i = A, B, C$$
$$v(A, B) = 2 \leq u'_A + u'_B$$
$$v(B, C) = 2 \leq u'_B + u'_C \tag{10.18}$$
$$v(A, C) = 2 \leq u'_A + u'_C$$
$$v(\mathfrak{A}) = 6 \leq u'_A + u'_B + u'_C$$

The above list shows that (i) unlike what happened in the benchmark set-up, all proper coalitions are equivalent, and (ii) the gain generated by the grand alliance is higher here than in the previous model, as each country is contiguous with the remaining two. The core is the region of the triangular world covering all the allocations undominated by any of the three proper coalitions. Moreover, the Shapley value is the triple $(2, 2, 2)$, reflecting again the full symmetry of the game.

10.3.2 *Coalition stability and the persistence of unipolarism*

Another interesting problem that has to do with alliances (and coalitions) is the issue of the persistence of unipolarism, which we have dealt with in Chapter 8. There, it was treated as a purely non-cooperative game. Here, we may take a further view at the same model from the standpoint of cooperative games, examining in particular the satellites' incentive to form a coalition.

To do so, let me recall the basic elements of the set-up. Players are one global power and two satellites. The *status quo ante* is that the global power is hegemonic, and the satellites contemplate the feasibility of a countervailing coalition. The interpretation of symbols is the same as in Chapter 8.

If the satellites form a coalition, each receives a payoff of

$$u_i(0, C) = c_i - k + C \qquad (10.19)$$

Otherwise, if satellite i receives a subsidy and this prevents the formation of the alliance, the payoff is

$$u_i(\sigma, NC) = c_i + \sigma \qquad (10.20)$$

At this point let me introduce some changes to the picture. First of all, define by $u_i(0, 0) = c_i$ the utility level of a satellite in the absence of both the coalition and the subsidy. Then, unlike what happens in the non-cooperative game, here there are two mutually exclusive perspectives: either the coalition forms (because either no subsidy is being paid or it is insufficient to avoid it) or not (because the subsidy has done its job). Moreover, the global power pays only a single subsidy to a single satellite for the sake of jeopardizing the alliance. Hence, the alternative perspectives for the global power are

$$U_G(\sigma, NC) = c_G - \sigma + \beta\sigma + P \qquad (10.21)$$

if a subsidy is paid and this prevents the formation of a coalition, or

$$U_G(0, C) = c_G + P - \beta C \qquad (10.22)$$

otherwise.

Now note that $u_i(0, 0)$ is strictly lower than both $u_i(0, C)$ and $u_i(\sigma, NC)$, so that there is an incentive to invest in the coalition, but there is also an incentive to accept the global power's subsidy. If the latter is weaker than the former, the coalition is unstable and the global power may reward a single satellite to attain its own goal, the persistence of unipolarism. This of course requires that paying a subsidy is convenient as opposed to letting the satellites build the coalition. The two conditions are

$$\sigma > C - k \qquad (10.23)$$

for the satellite, and

$$\sigma < \frac{\beta C}{1 - \beta} \qquad (10.24)$$

for the global power. These inequalities are compatible if

$$\frac{\beta C}{1 - \beta} > C - k \qquad (10.25)$$

or, equivalently, if

$$C(2\beta - 1) > k(\beta - 1) \qquad (10.26)$$

Since $\beta < 1$ by assumption, any $\beta \in (0, 1/2)$ is sufficient to ensure that the global power destabilizes the alliance by paying a subsidy corresponding to slightly more than $C - k$.

10.3.3   *Coalition stability and the environment*

As you surely know, the USA initially didn't sign the Kyoto Agreement. More recently, the COP15 meeting ended up with goodwill declarations that may not really deliver what seem to be – nominally – the intentions of the participants. Environmental policy is clearly a matter of agreements. Therefore, it is the ideal topic to be treated with the tools of bargaining and cooperative game theory. Here I am offering you a basic sketch of an environmental game where the players (countries) have to decide whether to join the coalition of those worrying about the planet (and investing to solve, or at least to mitigate, the planet's problems) or not to join and just free-ride on others' efforts. I will single out the conditions whereby one of these countries prefers not to join the coalition, leaving to the remaining two the burden of investing.

There are three fully symmetric (identical) countries, A, B and C (possibly the same as in the previous example, except that the particular shape of their borders is now irrelevant). They are aware that industrial activities produce a negative external effect e on the environment (e.g., pollution and global warming). In the absence of any investment to mitigate it, the payoff accruing to each country is a welfare level equal to

$$w(0) = y - e \tag{10.27}$$

where y is national income, the same for all. The zero in parentheses indicates that no country is investing.

There are three possible cases. The first is the 'one-man-band' coalition, with a single country – say, A – investing an amount of resources r to preserve the environment, and the remaining two free-riding upon it. In this case, payoffs are

$$w_A^i(1) = y - \frac{e}{k} - r$$
$$w_B^{fr}(1) = w_C^{fr}(1) = y - \frac{e}{k} \tag{10.28}$$

where superscripts i and fr stand for *investor* and *free-rider*, respectively. Parameter $k > 1$ measures the effectiveness of the investment r.

The second case is that of a proper coalition, with two countries – say, A and B – investing an amount k each. Here, payoffs are

$$w_A^i(2) = w_B^i(2) = y - \frac{e}{2k} - r$$
$$w_C^{fr}(1) = y - \frac{e}{2k} \tag{10.29}$$

The third and last case describes the grand coalition, with

$$w_A^i(3) = w_B^i(3) = w_C^i(3) = y - \frac{e}{3k} - r \tag{10.30}$$

Let's go step by step. First, we have to check whether there exists an individual incentive to invest when the remaining two countries, for sure, don't do it. This is the case provided that $w_A^i(1) > w(0)$, or equivalently

$$y - \frac{e}{k} - r \geq y - e \tag{10.31}$$

which implies that

$$r \le e\left(1 - \frac{1}{k}\right) \tag{10.32}$$

Second, we have to examine the individual incentive to join the proper coalition made up by two countries. The stability of this coalition requires two conditions: (i) a member should not want to abandon it to become a free-rider, and (ii) the outsider should prefer not to join it. That is, the following inequalities must be simultaneously satisfied:

$$y - \frac{e}{2k} - r \ge y - \frac{e}{k} \tag{10.33}$$

and

$$y - \frac{e}{2k} \ge y - \frac{e}{3k} - r \tag{10.34}$$

Condition (10.33) says that the payoff generated by the proper coalition must be higher than that generated by free-riding on the sole investor left after a unilateral deviation. Condition (10.34) says instead that the outsider must be better off by free-riding upon investors (rather than joining the coalition, which would thus become the *grand* one). Inequality (10.33) implies that

$$\frac{e}{2k} \ge r \tag{10.35}$$

while (10.33) is satisfied for all

$$r \ge \frac{e}{6k}$$

Hence, the proper coalition is stable for all

$$r \in \left[\frac{e}{6k}, \frac{e}{2k}\right] \tag{10.36}$$

That is, if r falls into the above interval, country C prefers not to join the coalition as it is ultimately convenient to free-ride on the latter's efforts.

Further reading

In addition to the references mentioned in the text, comprehensive overviews of bargaining and cooperation can be found in Aumann and Shapley (1974), Shubik (1982), Friedman (1986), Moulin (1988, 1995), Binmore (1992), Osborne and Rubinstein (1994), Montet and Serra (2003) and Rasmusen (2006). A detailed analysis of bargaining games can be found in Roth (1979); see also Kennan and Wilson (1993). On the experimental approach to bargaining, see Roth (1994). For more on other solution concepts for cooperative games, see Shapley (1962), Aumann and Drèze (1974), Maschler *et al.* (1979) and Roth (1988), *inter alia*. For an exposition of environmental games, see Dinar *et al.* (2009), Tisdell (2009) and Anderson (2010).

Notes

1 The origins: a bit of history

1 Indeed, the community of young game theorists working at Princeton in the early 1950s (Nash, Kuhn, etc.) themselves invented a parlour game they called 'Hex' (on this particular episode, see Nasar, 1998).

2 The discussion on how exactly one should interpret Zermelo's theorem is still going on. To my knowledge, the latest updating of this discussion is in Schwalbe and Walker (2001).

3 Additionally, von Neumann was indeed strongly anti-communist. In a famous interview released to *Life magazine*, when asked about the possibility of a first strike with nuclear weapons on the USSR, he said 'If you say why not bomb them tomorrow, I say why not today? If you say today at five o'clock, I say why not one o'clock?' (reported by Poundstone, 1992, p. 4). As we shall see in Chapter 3, this statement is very much reminiscent of backward induction.

4 To appreciate his figure and role in modern mathematics, consider that von Neumann solved one of the 23 problems appearing in Hilbert's programme. Moreover, his systematization of quantum mechanics is still a most elegant example of mathematical architecture (von Neumann, 1932). I'll say more about the relevance of this work and its connections with a few aspects of game theory in Chapter 3.

5 Later, von Neumann frankly admitted: 'without Oskar I would not have written that book' (as reported by Samuelson, 2001, p. 302).

6 In this and many other respects, the movie drastically differs from the book of the same title written by Sylvia Nasar (1998).

7 For a detailed account of this episode and von Neumann's reaction, see Nasar (1998, chapter 9).

8 A more thorough assessment should also include, at least, evolutionary game theory, which has an identity of its own and is of great relevance for biologists. Indeed, Maynard Smith (1982) has developed the concept of *evolutionary stable strategy* with the specific purpose of studying the strategic aspects of evolutionary behaviour in nature.

9 See, for instance, Schotter and Schwödiauer (1980, p. 480).

10 After the end of the Cold War, much more information has been made accessible to the general public. The websites of both institutions (www.rand.org and www.onr.navy.mil) report detailed accounts of the history and mission of both institutions, and what used to be confidential reports are now downloadable working papers.

11 Classical examples are Intriligator (1971), Başar and Olsder (1982), Mehlmann (1988) and Dockner *et al.* (2000).

12 By 'early sources' I mean the literature published between the end of World War II and the climax of the Cold War, say, the early 1980s.

13 Rufus Isaacs joined the RAND Corporation in autumn 1948. Then, starting from 1949, Isaacs and Bellman discussed new ideas about dynamic games and optimal control theory at RAND seminars, whose ultimate outcomes must be identified in Isaacs' *tenet of transition* and Bellman's *dynamic*

programming principle. In fact, the latter can be considered as a special case of the former. The same consideration applies to Pontryagin's *maximum principle*.

2 What is a game?

1 As well as to a large number of people – myself included – who have experienced similar situations over the last two decades.
2 In cases like this, the object is a *public good*. What makes it radically different from other non-public ones is the fact that it is *non-excludible* and *non-rival*.
3 A cautionary note: throughout the book, I will use the terms *action* and *strategy* as synonymous. Well, strictly speaking, they define different things, as one can quickly ascertain by consulting any (even slightly) more advanced textbook in game theory. An action is a *discrete strategy* (as in Matrices 2.1 and 2.2), while in general there are games played in *continuous strategies*, i.e., where players pick strategies from continuous intervals of real numbers. Yet, being more rigorous in this respect would hinder the readability of the book, while the level of approximation whereby action and strategy can be considered as perfect substitutes will do for my purposes.
4 To be more rigorous, one should define the players' utility functions, whose arguments can be those mentioned in the text, as well as many others. Then, players' behaviour would be aimed at the maximization of such utility functions.
5 Therefore, in the early days of game theory, it was also labelled as the Kuhn tree.
6 This is somewhat trivial in trees associated with 2×2 matrices, but certainly is not so in the uncountably many games that cannot be reduced to 2×2 matrices.

3 Solving a game

1 This is a very delicate aspect of game theory and economics in general. Nowadays, the same consideration also applies to most of the social sciences where an analogous approach is adopted (although it may not be unique or dominant). A thorough discussion of the common knowledge assumption would take us very far from the target of the present book, and therefore I will abstain from dwelling upon it any further. The interested reader may have a look at Binmore and Brandenbunger (1990) and Geanakoplos (1994).
2 As a useful exercise, one may verify that the outcome (s_{12}, s_{22}) is also selected by the minimax criterion, i.e., when each agent plays aggressively against the rival.
3 This term is named after the Italian scholar Vilfredo Pareto (1848–1923). The concept of Pareto efficiency can be briefly defined with reference to any situations involving an allocation of income, wealth or other resources, as follows. A Pareto-efficient allocation is that in which it is impossible to make an agent better off without necessarily making at least one other worse off.
4 The proof of the theorem is omitted as it is clearly beyond the scope of the present volume. The interested reader may find it in Fudenberg and Tirole (1991) and Osborne and Rubinstein (1994), *inter alia*.
5 Note that, although the district attorney is forced to interrogate the two players sequentially, still information is imperfect – as if the prisoners were interrogated simultaneously by two different attorneys – because they cannot communicate with each other.
6 Well, one could argue that the disaster that happens when she goes to the Superbowl and he goes to the movie theatre would yield largely negative payoffs. Although reasonable, this is inessential to our understanding of the game. Note that this outcome is not so unreasonable, as it can be thought to be the result of a purely altruistic attitude on the part of both players.
7 Again, also here there's a disastrous outcome, (c, c), giving rise to the least desirable payoffs as they cross the bridge and crash in the middle of it. However, as in the battle of the sexes, zeros will do.
8 That is, the game we are about to examine has been constructed on the basis of a well-known informal discussion about the arising of social conventions and rules in primitive societies, which can be found in Rousseau (1755).
9 That is, in games like this, there seems to be a recursive nature implicit in the coordination problem initially associated to the multiplicity of Nash equilibria: in trying to solve this coordination issue, players generate a similar one linked to the ambiguity in the order of deletion. This may look

like a purely philosophical problem, but this example shows that it may well have severe practical consequences.

10 There exists an alternative but entirely equivalent tree where player 2 is at the root while player 1 is at the intermediate nodes.

11 The games featuring this property are often labelled as 'Stackelberg-solvable games' (see d'Aspremont and Gérard-Varet, 1980), after the Austrian mathematical economist Heinrich von Stackelberg, who is recognized as the first to have investigated sequential play in non-cooperative games (Stackelberg, 1934).

12 See, for instance, McKelvey and Palfrey (1992), Nagel and Tang (1998) and Palacios-Huerta and Volij (2009).

13 See Güth *et al.* (1982), Güth (1995) and Gale *et al.* (1995), *inter alia.*

14 By the way, do you remember von Neumann's reply to the young Nash when the latter rushed into von Neumann's office back in 1949 (if you don't, go back to Chapter 1). The clockwise behaviour followed by players in the matching pennies game, as a result of unilateral deviations, indeed denotes the *lack of a fixed point* in the matrix.

15 For a contemporary revisitation of the dispute, see Hawking and Penrose (1996), with Hawking playing Bohr's role and Penrose playing Einstein's. A sketch of the equivalence between economics and physics when it comes to probabilities is given in the appendix to this chapter.

16 The material presented here is based on Lambertini (2000).

17 Erwin Schrödinger, together with Niels Bohr, Albert Einstein, Paul Dirac, Werner Heisenberg and Wolgang Pauli, is one of the founding fathers of quantum physics. The remainder of this section borrows from Hawking and Penrose (1996) and Penrose (1989, 1997).

18 This can be viewed as a *Gedankenexperiment* (thought experiment), in that we must not necessarily lock a cat into a box to verify the validity of what follows. In general, a *Gedankenexperiment* 'is consistent with the known laws of physics, even though it may not be technically feasible. So, measuring the acceleration due to gravity on the surface of the Sun is a *Gedankenexperiment*, whereas measuring the Doppler shift of sunlight as seen from a spaceship moving with twice the velocity of light is nonsense' (Gasiorowicz, 1996, p. 21, fn. 14). In the present case, one might replace *technically feasible* with *politically correct.*

19 von Neumann himself motivates his interest in quantum physics on the basis of his dissatisfaction with Dirac's formalization of the theory (see the author's preface in von Neumann (1932)).

20 Physicists keep having lively arguments concerning the alternative interpretation of such a statement, i.e., whether states *ca* and *cd* coexist in parallel universes or in the experimenter's mind. As far as the present book is concerned, this distinction, although intriguing, can be disregarded.

4 Understanding economics

1 Of course, a monopoly may not last forever, for several plausible reasons. For instance, some other firms may enter after having patented new technologies or products. I'll come to games dealing with this issue below.

2 I agree that it is much less than elegant to cite a work of one's own, but if you want to know more on the genesis of the so-called Bertrand model, see Lambertini and Mosca (2001) and references therein.

3 Remember that, in order to be appealing, a deviation must yield a strictly higher payoff to the deviator, as compared to what the same player gets by sticking to the candidate equilibrium outcome.

4 William Vickrey was the first to formalize the second-price auction (see Vickrey, 1962). He won the Nobel Prize in 1996, a few days before he died. The discipline investigating (*inter alia*) the formal structure of auctions is mechanism design, with a strong connection to game theory. In 2007, Leonid Hurwicz, Eric Maskin and Roger Myerson were awarded the Nobel Prize for their contribution to the optimal design of mechanisms.

5 In the jargon of industrial economics, this assumption says that the innovation is minor, or *non-drastic*. A drastic innovation grants the innovator monopoly power (with reference to the Bertrand game, any process innovation lowering the unit production cost c is automatically drastic).

6 In 2002, the patent laws in North America and the EU were harmonized, so that the duration of a patent on a major or drastic innovation is now 20 years on both sides of the Atlantic.

7 In this respect, an exemplifying story that many among you may remember is one that involved Airbus and Boeing in the late 1980s. The competition between Boeing and Airbus, in particular at its early stages, is a recurrent parable in any discussion about strategic trade policy involving the adoption of R&D subsidies – that helped Airbus enter the civil air transport industry, which of course was seen as unfair from the US standpoint (see Farnsworth, 1990; Krugman and Obstfeld, 1988; see also Krugman, 1987).

8 Note that, under the reverse sequence of play with the incumbent placed at the root, fighting would indeed be the subgame perfect outcome. Yet, a game where the incumbent fights entry before the entrant goes into the market is pure nonsense. The only sound interpretation of this game under sequential play is that depicted in Figure 4.2: if the outsider doesn't enter, there is no strategic interaction, which of course rules out a price war.

9 Recalling some of the *X-Files* we encountered in Chapter 1, it is fair to note that the internet is the by-product of Arpanet, which was first conceived for military use only.

10 Of course, this does not apply to cellular phones, which may be used to check email, take notes, shoot digital pictures, etc. In passing, it is worth mentioning the existence of *positional goods* yielding positional or *status* externalities. This is approximately the opposite of a network effect, since the positional externality decreases with the number of people able to access the same consumption pattern – if anybody can buy a yellow Porsche Carrera cabriolet, the latter ceases to be a status symbol.

11 For the sake of completeness, I should mention open-source software here. It is easy to understand that a thorough discussion of standardization and compatibility should also take into account R&D investments, the related patents establishing property rights, and their social desirability.

12 This effect is known as *switching cost*. For an overview of the theory of markets with switching costs, see Klemperer (1987, 1995).

13 Observe that inequality (4.14) is the condition that identifies a Nash equilibrium (you may go back to Chapter 3 to check this out).

14 This debate has generated such a large literature that it would be impossible to mention a fairly complete list of the most important contributions. I will confine myself to citing the core papers of Kydland and Prescott (1977) and Barro and Gordon (1983a, b).

15 This and the following game are loosely based on Downing and White (1989) and Milliman and Prince (1989).

16 This is clearly the case if R&D is fully subsidized, i.e., $\sigma = k$. This is, however, not strictly necessary to provide the entire industry with the appropriate incentive to invest in environmentally friendly technologies.

5 Repeated games and collusive behaviour

1 Note that, if the terminal date is T, then the game is repeated $T + 1$ times, as the initial date is $t = 0$.

2 One should take this assessment with a pinch of salt, as the asset in question might well be much more risky than government bonds. The recent financial crisis is there to prove beyond any reasonable doubt that this is not a remote possibility.

3 Again, in view of the recent financial crisis, the assumption of full refunding is not necessarily 100 per cent realistic.

4 The proof is beyond the scope of this book, and therefore I will skip it. However, for those familiar with Keynesian macroeconomics, it may be interesting to note that the mechanism generating (5.8) is the same one that yields the aggregate demand multiplier (see, e.g., Blanchard, 1997, chapter 3).

5 However, observe that experimental evidence points out that players tend to solve efficiently finitely repeated prisoners' dilemma games. See Dreber *et al.* (2008) and Rand *et al.* (2009), *inter alia*.

6 However, most of us believe that this should be attributed to Robert Aumann.

7 In this chapter, for the sake of simplicity, I will assume that there are no environmental externalities. For an interesting analysis of the relationship between collusion, environmental policy and R&D for green technologies, see Damania (1996).

8 Alternatively, one could envisage a deviation taking place at an arbitrary time τ during the supergame. This wouldn't make any difference, however. The reason is that, as long as the horizon is infinitely long, one may just relabel $\tau = 0$ and obtain exactly the same perspective.

9 This exercise is broadly based upon Majerus (1988), where a more general set-up allowing for the presence of product differentiation is considered.

10 A simplified version of such proof can be found in Rasmusen (2006, p. 112).
11 As the supergame lasts forever, if it is rational to fight entry at $t = 0$, then it is also rational to do it at any later date. As a result, if I adopts strategy f in the first period, firm E is correctly induced to believe that the incumbent will adopt it also at any point in the future.

6 Understanding politics

1 These requirements are *axioms*, i.e., that must be intuitively true and cannot be proven. We'll come back to the axiomatic approach in Chapter 10.
2 For the sake of being provocative, I could say that Hotelling – Bertrand = Downs. But I'm too serious to say anything like this in the main text. A simple note will do.
3 The reason for this assumption is to avoid perverse cases with the Left choosing a platform on the right-hand side of the preference spectrum (and conversely for the Right).
4 One could easily extend the model to allow for the presence of abstention. In this regard, see, for example, Hinich and Ordeshook (1970), Slutsky (1975) and Ordeshook (1992, chapter 3) and references therein.
5 An extension of the model specifically addressing this aspect of the model is laid out in the next section.
6 Again, this admittedly drastic assumption is adopted to keep things as simple as possible. A qualitatively analogous conclusion can be shown to hold under appropriate conditions in the more general case of asymmetric but still positive relocation costs, $C_L \neq C_R$.
7 Because the number of contributions dealing with this issue is extremely large, I will confine myself to redirect the interested reader to the beautiful overview contained in Persson and Tabellini (2000).
8 One could argue that the other party will itself make announcements to this regard. In order not to add any further complications, I am ruling out this possibility.
9 It is fair to say that this extensive form game is similar to others we have already seen earlier in the book (see, e.g., Figure 4.2).

7 Wargames

1 For further historical details on the battle of the Bismarck Sea, see Liddell Hart (1970, chapter 29).
2 Several detailed accounts of historical facts are available in the literature. See, for instance, Liddell Hart (1970), Keegan (1989) and Beevor (2009).
3 Escalations are also observed in other fields, where players overinvest in advertising, R&D or electoral campaigns (see Chapters 4 and 6). In such cases, however, the consequences are a great deal less harmful than in arms races.
4 For a revisitation of the dollar auction game in escalation models where the focus is explicitly on international relations, see O'Neill (1986) and Leininger (1989).
5 The following, strictly speaking, does not coincide with what is traditionally considered as the *MAD* game. I have adapted its structure to reflect the features of the Cuban missile crisis. For the original formulation of MAD, see, e.g., Gardner (1995, chapter 6).
6 To enrich our understanding of what exactly escalation means in this game, I may add that John F. Kennedy ordered (among other things) the Strategic Air Command bombers, armed with a gross total of 7000 megatons of nuclear weapons, to be kept aloft around the clock. For more details on this and other aspects, see Rhodes (1995).
7 This was actually a comprehensive attempt at selling an anodyne interpretation to the public in the West, but was essentially false. The neutron bomb, also known as *ERW* (*Enhanced Radiation Weapon*), would have caused environmental effects much worse than those that resulted from the nuclear incident that destroyed one of the Chernobyl reactors in April 1986.
8 I know, I know: all payoffs are disastrously negative (they measure terrible losses on both sides), so the idea of ranking outcomes according to the Pareto criterion sounds pretty much like a joke, but the decoupling strategy was also taking into account some idea of convenience that we may approximate here with Pareto efficiency.
9 As you may know, this measure is defined in terms of the equivalent amount of TNT. In itself, 50 kton means a terrible destructive power: approximately twice that of the atomic bombs that erased Hiroshima and Nagasaki in August 1945.
10 For a detailed account, see Fitzgerald (2001).

11 The choice of either strategy – in this case, whether or not to build a shield – may in fact signal an underlying attitude (aggressive or inoffensive, for instance). In this regard, an illuminating read is Fudenberg and Tirole (1984).

12 For the arising of the prisoners' dilemma structure in arms races, see Lichbach (1990).

13 All right, this is quite strong an assumption, leaving the initial set-up unmodified. For the analysis of the bearings of asymmetric effects exerted by a country's shield on the rival's payoffs, see Chassang and Padró i Miquel (2009).

14 For more on the concepts of risk and risk aversion in the theory of international relations, see O'Neill (2001).

8 Trade, security and hegemony

1 This may indeed sound like a justification of globalization. In fact, this view should be considered with a little caution, as it fits well the case of trade among countries with similar or at least comparable levels of income and consumer tastes (e.g., the EU, USA and Japan), while it is unable to describe the kind (and consequences) of trade between largely asymmetric countries, or more generally between the northern and southern hemispheres of the planet – in this regard, see Krugman (1979) and Flam and Helpman (1987).

2 For a wonderful account of the rise and fall of Great Britain as a sea power from the Tudors to the present day, see Kennedy (1976).

3 For further insights on these issues, see Snidal (1985), Yarbrough and Yarbrough (1986, 1987) and Powell (1991), *inter alia*.

4 The material presented here is based on Lambertini (2006).

5 In this regard, the existing literature is too wide to allow for an exhaustive listing, but see Morgenthau (1948), Waltz (1979) and Gilpin (1981), *inter alia*.

6 See also Ikenberry (1998/99, 2001) and Keohane (1989). Horowitz (2001) points out the relevance of appropriately accounting for risk-aversion, economic growth and political rigidities in assessing the performance of the balance-of-power theory.

7 Quoting Haass (1999, p. 37): 'trading some American power for a more stable international system would be a good deal for America and the world'.

8 A three-country world is also considered in some of the existing literature to analyse several issues. For instance, Wagner (2004) uses it to investigate the relation between bargaining and war.

9 The subsidy could, for example, consist of a pure money transfer or granting the satellites free access to G's domestic market, i.e., a free-trade agreement.

10 Note that the consumption level is not sensitive to the presence or absence of the subsidy. This will hold for the satellites as well. This entails that here I choose to abstract from any endogenous relationship between domestic welfare and international security, which relates to the 'guns or butter' debate.

11 This is acceptable insofar as the alliance is not a public good, i.e., *it takes two to tango*. The analysis of the game between satellites illustrating this feature of the model is given in the appendix to this chapter.

12 On the other hand, taking $\beta \in (0, 2/3)$ is necessary but by no means sufficient to ensure the opposite result, i.e., that $2k(1 - \beta) < C(2 - 3\beta)$. To see this, consider that, provided $2 - 3\beta > 0$, this inequality is equivalent to $C/(2k) > (1 - \beta)/(2 - 3\beta)$. Now, while we know that $C/(2k) > 1$ surely, in order for the coalition to be viable, it is also true that $(1 - \beta)/(2 - 3\beta) < 1$ for all $\beta \in (0, 1/2)$. Therefore, if $\beta \in (1/2, 2/3)$, then $(1 - \beta)/(2 - 3\beta) > 1$ so that $C/(2k)$ might be lower than $(1 - \beta)/(2 - 3\beta)$ in this range.

13 An alternative version of the one-shot set-up as a cooperative game is given in Chapter 10.

14 Alternatively, one could use Axelrod's (1981, 1984) *tit-for-tat* strategies, reaching qualitatively similar conclusions. See Chapter 5.

15 The objection could be raised that governments do not last indefinitely. However, the present analysis is designed with the view that they may be sufficiently forward-looking to closely mimic the behaviour of a single player living forever. Incidentally, this is in accordance with observation, at least with respect to the issue at hand.

16 Except, possibly, by invoking the focal point refinement (Schelling, 1960). See Chapter 3.

9 The role of information

1 The model is a simplified version of Mirrlees (1975, 1999), Holmström (1979), Shavell (1979) and Grossman and Hart (1983).
2 The idea that the principal can be taken to be risk-neutral, or indifferent, relies on the (untold) possibility for him/her to hold a portfolio of different assets (one being the firm we are looking at in this game) whose returns are reciprocally uncorrelated or even inversely correlated. Assuming risk-neutrality for the agent is, admittedly, a bit hard to swallow. For more on risk, uncertainty and portfolio theory, see Varian (1992).
3 Note that here the wage is not defined in terms of effort (which the principal cannot observe by assumption) but rather in terms of output level (the only thing that the principal observes, in addition, of course, to the market price and consequently profits).
4 This definition dates back, at least, to Nelson (1974).
5 Given that this would imply that the buyer doesn't know the type of the seller, this game could also be interpreted as one of incomplete information. Traditionally, however, it is considered as a prototypical game of asymmetric information.
6 If instead $\mathfrak{p}(T)\pi_w^E + [1 - \mathfrak{p}(T)]\pi_d > 0$ and $\pi_w^{WI} + \delta\pi_M > (1 + \delta)\pi_d$, the game has no Bayesian equilibrium in pure strategies. I leave this case out of the picture, as it is more complicated and ultimately inessential to the comprehension of the spirit of this game. If you are curious about the proof, see Montet and Serra (2003, p. 191).
7 If you want to have a look at the explicit expression by which the entrant computes the *ex post* probability of facing a tough incumbent, please see the appendix to this chapter.
8 This material is based on Tirole (1988, appendix to chapter 2).
9 This is a hedonic approach to consumer preferences that dates back to Mussa and Rosen (1978). A fine introduction to this topic can be found in Tirole (1988, chapter 2).
10 The dismissal of out-of-equilibrium beliefs involves the use of sequential rationality, yet another relevant refinement of the Nash equilibrium, which I am not explicitly dealing with here. See Kreps and Wilson (1982b) and Cho (1987), *inter alia*.
11 Should it decide to remain out of the 'nuclear club', country 1 would put itself at a disadvantage with respect to country 2, which might then be in a position to impose humiliating decisions on to country 1. This might explain the ranking of payoffs in the matrix.

10 Bargaining and cooperation

1 See Aumann (1985, 1987). You may also go back to Chapter 1.
2 There is more to it, although I'm well aware that by inserting any additional condition I may cause you to skip this part entirely. Collective rationality (10.9) also implies a feasibility condition $v(\mathfrak{C}) \geq \sum_{i=1}^N \pi_i$; however, this is not necessarily compatible with individual rationality (10.8), and the additional constraint $v(\mathfrak{C}) \geq \sum_{i=1}^N v_i(i)$ must be satisfied.
3 See Shapley (1967) and Owen (1968).
4 There are other 'values', although perhaps less relevant than Shapley's (see, e.g., Owen, 1968).
5 In some classes of games, additional axioms are also introduced. This is a delicate aspect that goes well beyond the scope of the present exposition. See Owen (1968) and Myerson (1991).
6 The symbol '!' means 'factorial'. In mathematics, the factorial of a positive integer n, denoted by $n!$, is the product of all positive integers less than or equal to n. For example, $3! = 1 \times 2 \times 3 = 6$. Note that $n!$ becomes very large rather quickly, as $4! = 1 \times 2 \times 3 \times 4 = 24$ but $5! = 1 \times 2 \times 3 \times 4 \times 5 = 120$.
7 Those of you interested to read more on this issue may see Olson and Zeckhauser (1966), Sandler and Cauley (1975), Sandler and Forbes (1980), Morrow (1991) and Hartley and Sandler (1999, 2001).

Bibliography

Abreu, D. J., 1986. 'Extremal Equilibria of Oligopolistic Supergames', *Journal of Economic Theory* 39: 191–225.

Abreu, D. J., 1988. 'On the Theory of Infinitely Repeated Games with Discounting', *Econometrica* 56: 383–96.

Adams, J., 1999. 'Policy Divergence in Multicandidate Probabilistic Spatial Voting', *Public Choice* 100: 103–22.

Adams, J., 2000. 'Multicandidate Equilibrium in American Elections', *Public Choice* 103: 297–325.

Akerlof, G., 1970. 'The Market for Lemons: Quality Uncertainty and the Market Mechanism', *Quarterly Journal of Economics* 84: 488–500.

Alchian, A. and Demsetz, H., 1972. 'Production, Information Costs and Economic Organization', *American Economic Review* 62: 777–95.

Anderson, D., 2010. *Environmental Economics and Natural Resource Management*, 3rd edn, Routledge, London.

Anderson, S., De Palma, A. and Thisse, J.-F., 1992. *Discrete Choice Theory of Product Differentiation*, MIT Press, Cambridge, MA.

Ansolabehere, S. and Snyder, J. M., Jr., 2000. 'Valence Politics and Equilibrium in Spatial Election Models', *Public Choice* 103: 327–36.

Arrow, K. J., 1951. *Social Choice and Individual Values*, Yale University Press, New Haven, CT.

Arrow, K. J. and Debreu, G., 1954. 'Existence of an Equilibrium for a Competitive Economy', *Econometrica* 22: 265–90.

Arrow, K. J., 1962. 'Economic Welfare and the Allocation of Resources for Invention', in Nelson, R., ed., *The Rate and Direction of Industrial Activity*, Princeton University Press, Princeton, NJ.

Arrow, K., 1985. 'The Economics of Agency', in Pratt, J. and Zeckhauser, R., eds, *Principals and Agents: The Structure of Business*, Harvard Business School Press, Boston, MA.

Aumann, R. J., 1964. 'Markets with a Continuum of Traders', *Econometrica* 32: 39–50.

Aumann, R. J., 1985. 'What Is Game Theory Trying to Accomplish?', in Arrow, K. and Honkapohja, S., eds, *Frontiers in Economics*, Blackwell, Oxford.

Aumann, R. J., 1987. 'Game Theory', in *The New Palgrave Dictionary of Economics*, Palgrave Macmillan, Basingstoke.

Aumann, R. J., 1999. 'Game Theory in Israel: Looking Backward and Forward', Plenary Lecture, ASSET Conference, Tel Aviv, October.

Aumann, R. J. and Drèze, J., 1974. 'Cooperative Games with Coalition Structures', *International Journal of Game Theory* 3: 217–37.

Aumann, R. J. and Shapley, L. S., 1974. *Values of Non-Atomic Games*, Princeton University Press, Princeton, NJ.

Austen-Smith, D., 1983. 'The spatial theory of electoral competition: instability, institutions, and information', *Environment and Planning C: Government and Policy* 1: 439–59.

Axelrod, R., 1981. 'The Emergence of Cooperation Among Egoists', *American Political Science Review* 75: 306–18.

Axelrod, R., 1984. *The Evolution of Cooperation*, Basic Books, New York.

Bain, J. 1956. *Barriers to New Competition*, Harvard University Press Cambridge, MA.

Banks, J. and Sobel, J., 1987. 'Equilibrium Selection in Signaling Games', *Econometrica* 55: 647–62.

Barro, R. J. and Gordon, D. B., 1983a. 'A Positive Theory of Monetary Policy in a Natural Rate Model', *Journal of Political Economy* 91: 589–610.

Barro, R. J. and Gordon, D. B., 1983b. 'Rules, Discretion and Reputation in a Model of Monetary Policy', *Journal of Monetary Economics* 12: 101–21.

Barro, R. J. and Sala-i-Martin, X., 1995. *Economic Growth*, McGraw-Hill, New York.

Bartholdi, J. J., Narasimhan, L. S. and Tovey, C. A., 1991. 'Recognizing Majority-Rule Equilibrium in Spatial Voting Games', *Social Choice and Welfare* 8: 183–97.

Başar, T. and Olsder, G. J., 1982. *Dynamic Noncooperative Game Theory*, Academic Press, San Diego, CA. [2nd edn, 1995.]

Beath, J. and Katsoulacos, Y., 1991. *The Economic Theory of Product Differentiation*, Cambridge University Press, Cambridge.

Beevor, A., 2009. *D-Day: The Battle for Normandy*, Penguin, London.

Belleflamme, P. and Peitz, M., 2010. *Industrial Organization: Markets and Strategies*, Cambridge University Press, Cambridge.

Berejikian, J. D., 2002. 'A Cognitive Theory of Deterrence', *Journal of Peace Research* 39: 165–83.

Bertrand, J., 1883. 'Review', *Journal des Savants*, 68: 499–508. [English translation available in Daughety, A., ed., *Cournot Oligopoly. Characterization and Applications*, pp. 73–81, Cambridge University Press, Cambridge, 1989.]

Binmore, K., 1992. *Fun and Games*, D. C. Heath, Lexington, MA.

Binmore, K. and Brandenbunger, A., 1990. 'Common Knowledge and Game Theory', in Binmore, K., ed., *Essays on the Foundations of Game Theory*, Blackwell, Oxford.

Binmore, K., Rubinstein, A. and Wolinsky, A., 1986. 'The Nash Bargaining Solution in Economic Modelling', *RAND Journal of Economics* 17: 176–88.

Binmore, K., Osborne, M. and Rubinstein, A., 1992. 'Non Cooperative Models of Bargaining', in Aumann, R. and Hart, S., eds, *Handbook of Game Theory with Economic Applications*, Vol. 1. North-Holland, Amsterdam.

Blanchard, O., 1997. *Macroeconomics*, Prentice-Hall, Englewood Cliffs, NJ.

Blanchard, O. and Fischer, S., 1989. *Lectures on Macroeconomics*, MIT Press, Cambridge, MA.

Bork, R. H., 1966. 'The Rule of Reason and the Per Se Concept: Price Fixing and Market Division', *Yale Law Journal* 75: 373–475.

Brams, S., 1975. *Game Theory and Politics*, Free Press, New York.

Brams, S., 1978. *The Presidential Elections Game*, Yale University Press, New Haven, CT.

Brandts, J. and Holt, C., 1992. 'An Experimental Test of Equilibrium Dominance in Signaling Games', *American Economic Review* 82: 1350–65.

Breitner, M. H., 2002. 'Rufus P. Isaacs and the Early Years of Differential Games: A Survey and Discussion Paper', in Petrosjan, L. A. and Zenkevich, N. A., eds, *Proceedings of the 10th International Symposium on Dynamic Games and Applications*, Vol. I, pp. 113–28, International Society of Dynamic Games and St. Petersburg State University.

Brito, D. L. and Intriligator, M. D., 1981. 'Strategic Arms Limitation Treaties and Innovations in Weapons Technology', *Public Choice* 37: 41–59.

Brito, D. L. and Intriligator, M. D., 1984. 'Can Arms Races Lead to the Outbreak of War?', *Journal of Conflict Resolution* 28: 63–84.

Brito, D. L. and Intriligator, M. D., 1994. 'Economic Aspects of Disarmament: Arms Race and Arms Control Issues', *Defence and Peace Economics* 5: 121–9.

Brito, D. L. and Intriligator, M. D., 1996. 'Proliferation and the Probability of War: A Cardinality Theorem', *Journal of Conflict Resolution* 40: 206–14.

Brito, D. L. and Intriligator, M. D., 2007. 'Arms Races and Proliferation', in Sandler, T. and Hartley, K., eds, *Handbook of Defense Economics. Defense in a Globalized World*, Elsevier, Amsterdam.

Brock, W. A. and Scheinkman, J. A., 1985. 'Price Setting Supergames with Capacity Constraints', *Review of Economic Studies* 52: 371–82.

Brouwer, L., 1910. 'Über eineindeutige, stetige Transformationen von Flächen in Sich', *Mathematische Annalen* 69: 176–80.

Cabral, L., 2000. *Introduction to Industrial Organization*, MIT Press, Cambridge, MA.

Calvert, R. L., 1985. 'Robustness of the Multidimensional Voting Model: Candidate Motivations, Uncertainty, and Convergence', *American Journal of Political Science* 29: 69–95.

Camera, G. and Casari, M., 2009. 'Cooperation Among Strangers Under the Shadow of the Future', *American Economic Review* 99: 979–1005.

Camerer, C., 2003. *Behavioral Game Theory. Experiments in Strategic Interaction*, Princeton University Press, Princeton, NJ.

Canzoneri, M. B., 1985. 'Monetary Policy Games and the Role of Private Information', *American Economic Review* 75: 1056–70.

Carrubba, C. J. and Singh, A., 2004. 'A Decision Theoretic Model of Public Opinion: Guns, Butter, and the European Common Defense', *American Journal of Political Science* 48: 218–31.

Caves, R. E. and Greene, D. P., 1996. 'Brands' Quality Levels, Prices, and Advertising Outlays: Empirical Evidence on Signals and Information Costs', *International Journal of Industrial Organization* 14: 29–52.

Chan, S. and Mintz, A., eds, 1992. *Defense, Welfare, and Growth*, Routledge, London.

Chang, M. H., 1991. 'The Effects of Product Differentiation on Collusive Pricing', *International Journal of Industrial Organization* 9: 453–69.

Chang, M. H., 1992. 'Intertemporal Product Choice and Its Effects on Collusive Firm Behavior', *International Economic Review* 33: 773–93.

Chassang, S. and Padró i Miquel, G., 2009. 'Defensive Weapons and Defensive Alliances', *American Economic Review* 99(2): 282–6.

Cho, I. K., 1987. 'A Refinement of Sequential Equilibrium', *Econometrica* 55: 1367–89.

Cho, I. K. and Kreps, D., 1987. 'Signaling Games and Stable Equilibria', *Quarterly Journal of Economics* 102: 179–221.

Conybeare, J. A. C., 1984. 'Public Goods, Prisoners' Dilemmas and the International Political Economy', *International Studies Quarterly* 28: 5–22.

Coughlin, P. J., 1990. 'Candidate Uncertainty and Electoral Equilibria', in Enelow, J. M. and Hinich, M. J., eds, *Advances in the Spatial Theory of Voting*, Cambridge University Press, Cambridge.

Coughlin, P. J., 1992. *Probabilistic Voting Theory*, Cambridge University Press, Cambridge.

Cournot, A., 1838. *Recherches sur les Principes Mathématiques de la Théorie des Richesses*, Hachette, Paris. [English translation: *Researches into the Mathematical Principles of the Theory of Wealth*, Macmillan, New York, 1897. The key chapter is also in Daugherty, A., ed., *Cournot Oligopoly. Characterization and Applications*, pp. 63–72, Cambridge University Press, Cambridge, 1989.]

Cox, M., 2005. 'Empire by Denial: The Strange Case of the United States', *International Affairs* 81: 15–30.

Dal Bo, P., 2005. 'Cooperation under the Shadow of the Future: Experimental Evidence from Infinitely Repeated Games', *American Economic Review* 95: 1591–604.

Damania, D., 1996. 'Pollution Taxes and Pollution Abatement in an Oligopoly Supergame', *Journal of Environmental Economics and Management* 30: 323–36.

d'Aspremont, C., Gabszewicz, J. J. and Thisse, J.-F., 1979. 'On Hotelling's "Stability in Competition" ', *Econometrica* 47: 1145–50.

d'Aspremont, C. and Gérard-Varet, L.-A., 1980. 'Stackelberg-Solvable Games', *Journal of Economic Theory* 23: 201–17; von Stackelberg, H., 1934. *Marktform und Gleichgewicht*. Springer-Verlag, Berlin.

Davis, O. A., DeGroot, M. H. and Hinich, M. J., 1972. 'Social Preference Orderings and Majority Rule', *Econometrica* 40: 147–57.

Deneckere, R., 1983. 'Duopoly Supergames with Product Differentiation', *Economics Letters* 11: 37–42.

Dinar, A., Albiac, J. and Sanchez-Soriano, J., 2009. *Game Theory and Policy Making in Natural Resources and the Environment*, Routledge, London.

Dirac, P. A. M., 1930. *The Principles of Quantum Mechanics*, Oxford University Press, Oxford. [4th edn, 1958.]

Dixit, A., 1979. 'A Model of Duopoly Suggesting a Theory of Entry Barriers', *Bell Journal of Economics* 10: 20–32.

Dixit, A. and Nalebuff, B., 1991. *Thinking Strategically. The Competitive Edge in Business, Politics, and Everyday Life*, Norton, New York.

Dockner, E. J., Jørgensen, S., Long, N. V. and Sorger, E. G., 2000. *Differential Games in Economics and Management Science*, Cambridge University Press, Cambridge.

Dorussen, H., 1999. 'Balance of Power Revisited: A Multi-Country Model of Trade and Conflict', *Journal of Peace Research* 36: 443–62.

Downing, P. B. and White, L. J., 1986. 'Innovation in Pollution Control', *Journal of Environmental Economics and Management* 13: 18–29.

Downs, A., 1957. *An Economic Theory of Democracy*, Harper and Row, New York.

Dreber, A., Rand, D., Ellingsen, T. and Nowak, M., 2008. 'Winners Don't Punish', *Nature* 452: 348–51.

Dunleavy, P., 1991. *Democracy, Bureaucracy and Public Choice: Economic Explanations in Political Science*, Prentice-Hall, Englewood Cliffs, NJ.

Dutta, P. K., 1999. *Strategies and Games*, MIT Press, Cambridge, MA.

Edgeworth, F. Y., 1881. *Mathematical Physics*, Kegan Paul, London.

Farnsworth, C. H., 1990. 'Aid to Airbus Called Unfair in US Study', *New York Times*, 8th September.

Fitzgerald, F., 2001. *Way Out There in the Blue: Reagan, Star Wars and the End of the Cold War*, Simon and Schuster, New York.

Flam, H. and Helpman, E., 1987. 'Vertical Product Differentiation and North–South Trade', *American Economic Review* 77: 810–22.

Flood, M., 1952. 'Some Experimental Games', RAND Report RM-798, RAND Corporation, Santa Monica, CA.

Flood, M., 1958. 'Some Experimental Games', *Management Science* 5: 5–26.

Friedman, J. W., 1971. 'A Non-Cooperative Equilibrium for Supergames', *Review of Economic Studies* 38: 1–12.

Friedman, J. W., 1986. *Game Theory with Applications to Economics*, Oxford University Press, Oxford.

Fudenberg, D. and Levine, D., 1998. *The Theory of Learning in Games*, MIT Press, Cambridge, MA.

Fudenberg, D. and Maskin, E., 1986. 'The Folk Theorem in Repeated Games with Discounting or with Incomplete Information', *Econometrica* 54: 533–54.

Fudenberg, D. and Tirole, J., 1984. 'The Fat-Cat Effect, the Puppy-Dog Ploy, and the Lean and Hungry Look', *American Economic Review* 74(2): 361–6.

Fudenberg, D. and Tirole, J., 1991. *Game Theory*, MIT Press, Cambridge, MA.

Gabszewicz, J. J. and Vial, J.-P., 1972. 'Oligopoly *à la* Cournot in a General Equilibrium Analysis', *Journal of Economic Theory* 4: 381–400.

Gaddis, J. L., 2006. *The Cold War: A New History*, Pengiun Press, New York.

Gale, J., Binmore, K. G. and Samuelson, L., 1995. 'Learning to be Imperfect: the Ultimatum Game', *Games and Economic Behavior* 8: 56–90.

Gallini, N., 1992. 'Patent Policy and Costly Imitation', *RAND Journal of Economics* 23: 52–63.

Gardner, R., 1995. *Games for Business and Economics*, Wiley, New York.

Gasiorowicz, S., 1996. *Quantum Physics*, 2nd edn, Wiley, New York.

Geanakoplos, J., 1994. 'Common Knowledge', in Aumann, R. and Hart, S., eds, *Handbook of Game Theory with Economic Applications*, Vol. 2, North-Holland, Amsterdam.

Gibbons, R., 1992. *A Primer in Game Theory*, Harvester-Wheatsheaf, Englewood Cliffs, NJ.

Gilbert, R. and Newbery, D., 1982. 'Preemptive Patenting and the Persistence of Monopoly', *American Economic Review* 72: 514–26.

Gillies, D. B., 1959. 'Solutions to General Non-Zero-Sum Games', in Tucker, A. W. and Luce, D. R., eds, *Contributions to the Theory of Games*, Vol. IV, Annals of Mathematical Studies, No. 40, Princeton University Press, Princeton, NJ.

Gilpin, R., 1981. *War and Change in World Politics*, Princeton University Press, Princeton, NJ.

Goeree, J. K. and Holt, C. A., 2001. 'Ten Little Treasures of Game Theory and Ten Intuitive Contradictions', *American Economic Review* 91: 1402–22.

Gowa, J., 1989. 'Bipolarity, Multipolarity, and Free Trade', *American Political Science Review* 83: 1245–56.

Gowa, J., 1994. *Allies, Adversaries, and International Trade*, Princeton University Press, Princeton, NJ.

Green, E. and Porter, R., 1984. 'Non-Cooperative Collusion Under Imperfect Price Information', *Econometrica* 52: 87–100.

Grossman, G. M., ed., 1992. *Imperfect Competition and International Trade*, MIT Press, Cambridge, MA.

Grossman, S. and Hart, O., 1983. 'An Analysis of the Principal-Agent Problem', *Econometrica* 50: 7–45.

Güth, W., 1995. 'On Ultimatum Bargaining Experiments – A Personal Review', *Journal of Economic Behavior and Organization* 27: 329–44.

Güth, W., Schmittberger, R. and Schwarze, B., 1982. 'An Experimental Analysis of Ultimatum Bargaining', *Journal of Economic Behavior and Organization* 3: 367–88.

Haas, M. L., 2001. 'Prospect Theory and the Cuban Missile Crisis', *International Studies Quarterly* 45: 241–70.

Haass, R. N., 1999. 'What to Do with American Primacy', *Foreign Affairs* 78: 37–49.

Hardin, G., 1968. 'The Tragedy of the Commons', *Science* 162: 1243–8.

Harsanyi, J., 1967/68. 'Games with Incomplete Information Played by Bayesian Players, Parts I, II and III', *Management Science* 14: 159–82, 320–34, 486–502.

Harsanyi, J., 1977. *Rational Behavior and Bargaining Equilibrium in Games and Social Situations*, Cambridge University Press, Cambridge.

Harsanyi, J. and Selten, R., 1972. 'A Generalized Nash Solution for Two-Person Bargaining Games with Incomplete Information', *Management Science* 18: 80–106.

Harsanyi, J. and Selten, R., 1988. *A General Theory of Equilibrium Selection in Games*, MIT Press, Cambridge, MA.

Hartley, K. and Sandler, T., 1999. 'NATO Burden Sharing: Past and Future', *Journal of Peace Research* 36: 665–80.

Hartley, K. and Sandler, T., 2001. 'Economics of Alliances: The Lessons for Collective Action', *Journal of Economic Literature* 39: 869–96.

Hauk, E., 2003. 'Multiple Prisoner's Dilemma Games With(out) an Outside Option: An Experimental Study', *Theory and Decision* 54: 207–29.

Hawking, S. and Penrose, R., 1996. *The Nature of Space and Time*, Princeton University Press, Princeton, NJ.

Haywood, O. G., Jr., 1954. 'Military Decision and Game Theory', *Journal of the Operations Research Society of America* 2: 365–85.

Helpman, E. and Krugman, P., 1985. *Market Structure and Foreign Trade*, MIT Press, Cambridge, MA.

Helpman, E. and Krugman, P., 1989. *Trade Policy and Market Structure*, MIT Press, Cambridge, MA.

Heo, U., 1998. 'Modeling the Defense–Growth Relationship Around the Globe', *Journal of Conflict Resolution* 42: 637–57.

Hinich, M. J., 1977. 'Equilibrium in Spatial Voting: The Median Voter Result is an Artifact', *Journal of Economic Theory* 16: 208–19.

Hinich, M. J. and Ordeshook, P., 1970. 'Plurality Maximization vs. Vote Maximization: A Spatial Analysis with Variable Participation', *American Political Science Review* 64: 772–91.

Hinich, M. J., Ledyard, J. D. and Ordeshook, P., 1972. 'Nonvoting and the Existence of Equilibrium Under Majority Rule', *Journal of Economic Theory* 4: 44–53.

Hinich, M. J., Ledyard, J. D. and Ordeshook, P., 1973. 'A Theory of Electoral Equilibrium: A Spatial Analysis Based on the Theory of Games', *Journal of Politics* 35: 154–93.

Holmström, B., 1979. 'Moral Hazard and Observability', *Bell Journal of Economics* 10: 74–91.

Holmström, B., 1982. 'Moral Hazard in Teams', *Bell Journal of Economics* 13: 324–40.

Horowitz, S., 2001. 'The Balance of Power: Formal Perfection and Practical Flaws', *Journal of Peace Research* 38: 705–22.

Hotelling, H., 1929. 'Stability in Competition', *Economic Journal* 39: 41–57.

Huck, S., Knoblauch, V. and Muller, W., 2006. 'Spatial Voting with Endogenous Timing', *Journal of Institutional and Theoretical Economics* 162: 557–70.

Ikenberry, G. J., 1998/99. 'Institutions, Strategic Restraint, and the Persistence of American Postwar Order', *International Security* 23: 43–78.

Ikenberry, G. J., 2001. *After Victory*, Pricenton University Press, Princeton, NJ.

Ikenberry, G. J., ed., 2002. *America Unrivaled. The Future of the Balance of Power*, Cornell University Press, Ithaca, NY.

Ikenberry, G. J., 2004. 'Liberalism and Empire: Logics of Order in the American Unipolar Age', *Review of International Studies* 30: 609–30.

Intriligator, M. D., 1971. *Dynamic Optimization and Economic Theory*, Prentice-Hall, Englewood Cliffs, NJ.

Isaacs, R., 1951. 'Games of Pursuit', RAND Discussion Paper P-257, RAND Corporation, Santa Monica, CA.

Isaacs, R., 1965. *Differential Games*, Wiley, New York.

Isaacs, R., 1973. 'Differential Games: Their Scope, Nature and Future', in Blaquiere, A., ed., *Topics in Differential Games*, pp. 1–42, North-Holland, Amsterdam.

Iyori, H. and Uesugi, A., 1983. *The Antimonopoly Law of Japan*, Federal Legal Publications.

Kakutani, S., 1941. 'A Generalization of Brouwer's Fixed Point Theorem', *Duke Mathematical Journal* 8: 457–9.

Kalai, E., 1977. 'Proportional Solutions to Bargaining Situations: Intertemporal Utility Comparisons', *Econometrica* 45, 1623–30.

Kalai, E. and Smorodinsky, M., 1975. 'Other Solutions to Nash's Bargaining Problem', *Econometica* 43: 513–18.

Kalai, E., Samet, D. and Stanford, W., 1988. 'A Note on Reactive Equilibria in the Discounted Prisoners' Dilemma', *International Journal of Game Theory* 3: 177–86.

Keegan, J., 1989. *The Second World War*, Hutchinson, London.

Kennan, J. and Wilson, R., 1993. 'Bargaining with Private Information', *Journal of Economic Literature* 31: 45–104.

Kennedy, P., 1976. *The Rise and Fall of British Naval Mastery*, Penguin, London.

Keohane, R. O., 1986. 'Reciprocity in International Relations', *International Organization* 40: 1–27.

Keohane, R. O., 1989. *International Institutions and State Power*, Westview Press, Boulder, CO.

Kihlstrom, R. E. and Riordan, M. H., 1984. 'Advertising as a Signal', *Journal of Political Economy* 92: 427–50.

Kilgour, D. M. and Zagare, F. C., 1991. 'Credibility, Uncertainty, and Deterrence', *American Journal of Political Science* 35: 305–34.

Klein, B. and Leffler, K., 1981. 'The Role of Market Forces in Assuring the Contractual Performance', *Journal of Political Economy* 89: 615–41.

Klemperer, P., 1987. 'Markets with Consumer Switching Costs', *Quarterly Journal of Economics* 102: 375–94.

Klemperer, P., 1995. 'Competition When Consumers Have Switching Costs', *Review of Economic Studies* 62: 515–39.

Klemperer, P., 2003. 'Why Every Economist Should Learn Some Auction Theory', in Dewatripont, M., Hansen, L. and Turnovsky, S., eds, *Advances in Economics and Econometrics: Invited Lectures to 8th World Congress of the Econometric Society*, pp. 25–55, Cambridge University Press, Cambridge. [A downloadable version can be found on Paul Klemperer's home page, http://www.gqq10.dial.pipex.com/.]

Klemperer, P., 2004. *Auctions. Theory and Practice*, Princeton University Press, Princeton, NJ.

Kohlberg, E. and Mertens, J.-F., 1986. 'On the Strategic Stability of Equilibria', *Econometrica* 54: 1003–38.

Kramer, G. H., 1978. 'Existence of Electoral Equilibrium', in Ordeshook, P., ed., *Game Theory and Political Science*, New York University Press, New York.

Kreps, D. and Scheinkman, J., 1983. 'Quantity precommitment and bertrand competition yield cournot outcomes', *Bell Journal of Economics* 14: 326–37.

Kreps, D. and Wilson, R., 1982a. 'Reputation and Imperfect Information', *Journal of Economic Theory* 27: 253–79.

Kreps, D. and Wilson, R., 1982b. 'Sequential Equilibrium', *Econometrica* 50: 863–94.

Krishna, V., 2002. *Auction Theory*, Academic Press, San Diego, CA.

Krishna, V. and Morgan, J., 1997. 'An Analysis of the War of Attrition and the All-Pay Auction', *Journal of Economic Theory* 72: 343–62.

Krugman, P., 1979. 'A Model of Innovation, Technology Transfer, and the World Distribution of Income', *Journal of Political Economy* 87: 253–66.

Krugman, P., 1987. 'Is Free Trade Passé?', *Journal of Economic Perspectives* 1: 131–44.

Krugman, P., 1990. *Rethinking International Trade*, MIT Press, Cambridge, MA.

Krugman, P., 1991. *Geography and Trade*, MIT Press, Cambridge, MA.

Krugman, P. and Obstfeld, M., 1988. *International Economics*, Scott, Foresman & Co., Glenview, IL.

Kuhn, H. W., 1953. 'Extensive Games and the Problem of Information', in Kuhn, H. W. and Tucker, A. W., eds, *Contributions to the Theory of Games*, Vol. II, Annals of Mathematical Studies, No. 28, Princeton University Press, Princeton, NJ.

Kupchan, C., 2002. *The End of the American Era*, Knopf, New York.

Kydd, A., 2000. 'Arms Races and Arms Control: Modeling the Hawk Perspective', *American Journal of Political Science* 44: 228–44.

Kydland, F. and Prescott, E., 1977. 'Rules Rather Than Discretion: The Inconsistency of Optimal Plans', *Journal of Political Economy* 85: 473–92.

Lambertini, L., 2000. 'Quantum Mechanics and Mathematical Economics Are Isomorphic. John von Neumann between Physics and Economics', Working Paper 370, Department of Economics, University of Bologna. [Downloadable at http://www2.dse.unibo.it/lamberti/johnvn.pdf.]

Lambertini, L., 2006. 'Is America Unrivaled? A Repeated Game Analysis', Working Paper 563, Department of Economics, University of Bologna. [Downloadable at http://www2.dse. unibo.it/wp/563.pdf.]

Lambertini, L., 2007. 'Platform Stickiness in a Spatial Voting Model', *Economics Bulletin* 4: 1–11.

Lambertini, L. and Mosca, M., 2001. 'Give to Caesar What Is Caesar's. Or, Give to Launhardt What We Are Used to Think Is Bertrand's', Working Paper 413, Dipartimento di Scienze Economiche, Università degli Studi di Bologna. [Dowloadable at http:// www2.dse.unibo.it/lamberti/hoperev.pdf.]

Lebow, R. N. and Gross Stein, J., 1989. 'Rational Deterrence Theory: I Think, Therefore I Deter', *World Politics* 41: 208–24.

Leininger, W., 1989. 'Escalation and Cooperation in Conflict Situations: The Dollar Auction Revisited', *Journal of Conflict Resolution* 33: 231–54.

Leonard, R. J., 1994. 'Reading Cournot, Reading Nash: The Creation and Stabilisation of the Nash Equilibrium', *Economic Journal* 104: 492–511.

Leonard, R. J., 1995. 'From Parlor Games to Social Science: von Neumann, Morgenstern, and the Creation of Game Theory', *Journal of Economic Literature* 33: 730–61.

Lichbach, M. I., 1990. 'When is an Arms Rivalry a Prisoner's Dilemma? Richardson's Models and 2×2 Games', *Journal of Conflict Resolution* 34: 29–56.

Liddell Hart, B. H., 1970. *History of the Second World War*, Cassell, London.

Luce, R. D. and Raiffa, H., 1957. *Games and Decisions. Introduction and Critical Survey*, Wiley, New York.

Majerus, D., 1988. 'Price vs Quantity Competition in Oligopoly Supergames', *Economics Letters* 27: 293–7.

Martin, S., 1993. *Advanced Industrial Economics*, Blackwell, Oxford.

Martin, S., ed., 1998. *Competition Policies in Europe*, North-Holland, Amsterdam.

Martin, S., 2002. *Advanced Industrial Economics, Second Edition*, Blackwell, Oxford.

Martin, S., 2004. 'Remembrance of Things Past', Unpublished Manuscript, Purdue University, West Lafayette, IN.

Maschler, M., Peleg, B. and Shapley, L. S., 1979. 'Geometric Properties of the Kernel, Nucleolus, and Related Solution Concepts', *Mathematics of Operations Research* 4: 303–38.

Mastanduno, M., 1997. 'Preserving the Unipolar Moment: Realist Theories and U.S. Grand Strategy After the Cold War', *International Security* 21: 49–88.

Maynard Smith, J., 1982. *Evolution and the Theory of Games*, Cambridge University Press, Cambridge.

McKelvey, R. D. and Ordeshook, P., 1976. 'Symmetric Spatial Games Without Majority Rule Equilibria', *American Political Science Review* 70: 1172–84.

McKelvey, R. and Palfrey, T., 1992. 'An Experimental Study of the Centipede Game', *Econometrica* 60: 803–36.

McKelvey, R. D. and Patty, J. W., 2006. 'A Theory of Voting in Large Elections', *Games and Economic Behavior* 57: 155–80.

Mehlmann, A., 1988. *Applied Differential Games*, Plenum Press, New York.

Mertens, J.-F. and Zamir, S., 1985. 'Formulation of Bayesian Analysis for Games with Incomplete Information', *International Journal of Game Theory* 14: 1–29.

Milgrom, P. and Roberts, J., 1982a. 'Predation, Reputation, and Entry Deterrence', *Journal of Economic Theory* 27: 280–312.

Milgrom, P. and Roberts, J., 1982b. 'Limit Pricing and Entry Under Incomplete Information', *Econometrica* 50: 443–60.

Milgrom, P. and Roberts, J., 1986. 'Price and Advertising Signals of Product Quality', *Journal of Political Economy* 94: 796–821.

Milliman, S. R. and Prince, R., 1989. 'Firm Incentives to Promote Technological Change in Pollution Control', *Journal of Environmental Economics and Management* 17: 247–65.

Mirowski, P., 2002. *Machine Dreams. Economics Becomes a Cyborg Science*, Cambridge University Press, Cambridge.

Mirrlees, J., 1975. 'The Theory of Moral Hazard and Unobservable Behaviour, Part I', Mimeo, Nuffield College, University of Oxford.

Mirrlees, J., 1999. 'The Theory of Moral Hazard and Unobservable Behaviour: Part I', *Review of Economic Studies* 66: 3–21.

Montet, C. and Serra, D., 2003. *Game Theory and Economics*, Palgrave Macmillan, Basingstoke.

Morgenthau, H., 1948. *Politics Among Nations*, Knopf, New York.

Morrow, J. D., 1986. 'A Spatial Model of International Conflict', *American Political Science Review* 80: 1131–50.

Morrow, J. D., 1991. 'Alliances and Asymmetry: An Alternative to Capability Aggregation Model of Alliances', *American Journal of Political Science* 35: 904–13.

Morrow, J. D., 1992. *Game Theory for Political Scientists*, Princeton University Press, Princeton, NJ.

Moulin, H., 1988. *Axioms of Cooperative Decision Making*, Cambridge University Press, Cambridge.

Moulin, H., 1995. 'An Appraisal of Cooperative Game Theory', *Revue d'Économie Politique* 105: 617–32.

Mussa, M. and Rosen, S., 1978. 'Monopoly and Product Quality', *Journal of Economic Theory* 18: 301–17.

Muthoo, A., 1999. *Bargaining Theory with Applications*, Cambridge University Press, Cambridge.

Myerson, R., 1991. *Game Theory. Analysis of Conflict*, Harvard University Press, Cambridge, MA.

Myerson, R., 1999. 'Nash Equilibrium and the History of Economic Theory', *Journal of Economic Literature* 37: 1067–82.

Nagel, R. and Tang, F. F., 1998. 'An Experimental Study on the Centipede Game in Normal Form – An Investigation on Learning', *Journal of Mathematical Psychology* 42: 356–84.

Nalebuff, B., 1991. 'Rational Deterrence in an Imperfect World', *World Politics* 43: 313–35.

Nasar, S., 1998. *A Beautiful Mind*, Touchstone Books, New York.

Nash, J. F., 1950a. 'Equilibrium Points in n-Person Games', *Proceedings of the National Academy of Science of the USA* 36: 48–9.

Nash, J. F., 1950b. 'The Bargaining Problem', *Econometrica* 18: 155–62.

Nash, J. F., 1951. 'Non-Cooperative Games', *Annals of Mathematics* 54: 289–95.

Nash, J. F., 1953. 'Two Person Cooperative Games', *Econometrica* 21: 128–40.

Nelson, P., 1974. 'Advertising as Information', *Journal of Political Economy* 82: 729–54.

Olson, M. and Zeckhauser, R., 1966. 'An Economic Theory of Alliances', *Review of Economics and Statistics* 48: 266–79.

O'Neill, B., 1986. 'International Escalation and the Dollar Auction', *Journal of Conflict Resolution* 30: 33–50.

O'Neill, B., 2001. 'Risk Aversion in International Relations Theory', *International Studies Quarterly* 45: 617–40.

Ordeshook, P., 1986. *Game Theory and Political Theory*, Cambridge University Press, Cambridge.

Ordeshook, P., 1992. *A Political Theory Primer*, Routledge, London.

Osborne, M. J., 1995. 'Spatial Models of Political Competition under Plurality Rule: A Survey of Some Explanations of the Number of Candidates and the Positions They Take', *Canadian Journal of Economics* 28: 261–301.

Osborne, M. J., 2003. *An Introduction to Game Theory*, Oxford University Press, Oxford.

Osborne, M. J. and Rubinstein, A., 1994. *A Course in Game Theory*, MIT Press, Cambridge, MA.

Owen, G., 1968. *Game Theory*, Academic Press, New York. [3rd edn, 1995.]

Palacios-Huerta, I. and Volij, O., 2009. 'Field Centipedes', *American Economic Review* 99: 1619–35.

Patty, J. W., 2005. 'Local Equilibrium Equivalence in Probabilistic Voting Models', *Games and Economic Behavior* 51: 523–36.

Pearce, D. W. and Turner, R. K., 1989. *Economics of Natural Resources and the Environment*, Harvester-Wheatsheaf, Hemel Hempstead.

Penrose, R., 1989. *The Emperor's New Mind*, Oxford University Press, Oxford.

Penrose, R., 1997. *The Large, the Small and the Human Mind*, Cambridge University Press, Cambridge.

Persson, T. and Tabellini, G., 1990. *Macroeconomic Policy, Credibility, and Politics*, Harwood Academic, London.

Persson, T. and Tabellini, G., 2000. *Political Economics. Explaining Economic Policy*, MIT Press, Cambridge, MA.

Polachek, S. W., 1980. 'Conflict and Trade', *Journal of Conflict Resolution* 24: 55–78.

Pontryagin, L. S., 1966. 'On the Theory of Differential Games', *Uspekhi Matematicheskikh Nauk* 21: 219–74.

Poundstone, E., 1992. *Prisoner's Dilemma. John von Neumann, Game Theory, and the Puzzle of the Bomb*, Anchor Books, New York.

Powell, R., 1987. 'Crisis Bargaining, Escalation, and MAD', *American Political Science Review* 81: 717–36.

Powell, R., 1989a. 'Crisis Stability in the Nuclear Age', *American Political Science Review* 83: 61–76.

Powell, R., 1989b. 'Nuclear Deterrence and the Strategy of Limited Retaliation', *American Political Science Review* 83: 503–19.

Powell, R., 1990. *Nuclear Deterrence Theory. The Search for Credibility*, Cambridge University Press, Cambridge.

Powell, R., 1991. 'Absolute and Relative Gains in International Relations Theory', *American Political Science Review* 85: 1303–20.

Powell, R., 1993. 'Guns, Butter, and Anarchy', *American Political Science Review* 87: 115–32.

Powell, R., 2003. 'Nuclear Deterrence Theory, Nuclear Proliferation, and National Missile Defense', *International Security* 27: 86–118.

Ramsey, F. P., 1928. 'A Mathematical Theory of Saving', *Economic Journal* 38: 543–9. [Reprinted in Stiglitz, J. E. and Uzawa, H., eds, *Readings in the Modern Theory of Economic Growth*, MIT Press, Cambridge, MA, 1969.]

Ramsey, F. P., 1931. 'Truth and Probability (1926)', in Braithwaite, R. B., ed., *The Foundations of Mathematics and Other Logical Essays*, Routledge and Kegan Paul, London. [Humanities Press, New York, 1950.]

Rand, D., Dreber, A., Ellingsen, T., Fudenberg, D. and Nowak, M., 2009. 'Positive Interactions Promote Public Cooperation', *Science* 325: 1272–5.

Rasmusen, E., 2006. *Games and Information: An Introduction to Game Theory*, 4th edn, Blackwell, Oxford.

Reinganum, J., 1983. 'Uncertain Innovation and the Persistence of Monopoly', *American Economic Review* 73: 741–8.

Rhodes, R., 1995. *Dark Sun. The Making of the Hydrogen Bomb*, Simon and Schuster, New York.

Roemer, J. E., 1994. 'A Theory of Policy Differentiation in Single Issue Electoral Politics', *Social Choice and Welfare* 11: 355–80.

Rosenthal, R. W., 1981. 'Games of Perfect Information, Predatory Pricing and the Chain-Store Paradox', *Journal of Economic Theory* 25: 92–100.

Ross, T. W., 1992. 'Cartel Stability and Product Differentiation', *International Journal of Industrial Organization* 10: 1–13.

Rotemberg, J. and Saloner, G., 1986. 'A Supergame-Theoretic Model of Price Wars During Booms', *American Economic Review* 76: 390–407.

Roth, A. E., 1979. *Axiomatic Models of Bargaining*, Lecture Notes in Economics and Mathematical Systems, Vol. 170, Springer, Heidelberg.

Roth, A. E., ed., 1988. *The Shapley Value. Essays in Honor of Lloyd S. Shapley*, Cambridge University Press, Cambridge.

Roth, A. E., 1993. 'On the Early History of Experimental Economics', *Journal of the History of Economic Thought* 15: 184–209.

Roth, A. E., 1994. 'Bargaining Experiments', in Kogel, J. H. and Roth, A. E., eds, *Handbook of Experimental Economics*, North-Holland, Amsterdam.

Rothschild, M. and Stiglitz, J., 1976. 'Equilibrium in Competitive Insurance Markets: An Essay on the Economics of Imperfect Information', *Quarterly Journal of Economics* 90: 629–49.

Rousseau, J.-J., 1755. *Discours sur l'Origine et les Fondements de l'Inégalité Parmi les Hommes*.

Rubinstein, A., 1982. 'Perfect Equilibrium in a Bargaining Model', *Econometrica* 50: 97–109.

Sabater-Grande. G. and Georgantzis, N., 2002. 'Accounting for Risk Aversion in Repeated Prisoners' Dilemma Games: An Experimental Test', *Journal of Economic Behavior and Organization* 48: 37–50.

Samuelson, P. A., 2001. 'Some Game Theory Anecdotes', *Japan and the World Economy* 13: 299–302.

Sandler, T., 1999. 'Alliance Formation, Alliance Expansion, and the Core', *Journal of Conflict Resolution* 43: 727–47.

Sandler, T. and Cauley, J., 1975. 'On the Economic Theory of Alliances', *Journal of Conflict Resolution* 19: 330–48.

Sandler, T. and Forbes, J. F., 1980. 'Burden Sharing, Strategy, and the Design of NATO', *Economic Inquiry* 18: 425–44.

Schelling, T., 1960. *The Strategy of Conflict*, Harvard University Press, Cambridge, MA.

Schmalensee, R., 1978. 'A Model of Advertising and Product Quality', *Journal of Political Economy* 86: 485–503.

Schofield, N., 2006. 'Equilibria in the Spatial Stochastic Model of Voting with Party Activists', *Review of Economic Design* 10: 183–203.

Schotter, A. and Schwödiauer G., 1980. 'Economics and the Theory of Games: A Survey', *Journal of Economic Literature* 18: 479–527.

Schumpeter, J. A., 1934. *The Theory of Economic Development*, Harvard University Press, Cambridge, MA.

Schumpeter, J. A., 1942. *Capitalism, Socialism, and Democracy*, Harper, New York.

Schwalbe, U. and Walker, P., 2001. 'Zermelo and the Early History of Game Theory', *Games and Economic Behavior* 34: 123–37.

Scotchmer, S., 2004. *Innovation and Incentives*, MIT Press, Cambridge, MA.

Selten, R., 1965. 'Spielteoretische Behandlung eines Oligopolmodells mit Nachfrageträgheit', *Zeitschrift für die gesamte Staatswissenschaft* 121: 301–24.

Selten, R., 1975. 'Re-Examination of the Perfectness Concept for Equilibrium Points in Extensive Games', *International Journal of Game Theory* 4: 25–55.

Selten, R., 1978. 'The Chain-Store Paradox', *Theory and Decision* 9: 127–59.

Shapiro, C., 1983. 'Premiums for High Quality Product as Returns to Reputation', *Quarterly Journal of Economics* 98: 659–79.

Shapley, L. S., 1953. 'A Value for N-Person Games', in Kuhn, H. W. and Tucker, A. W., eds, *Contributions to the Theory of Games*, Vol. II, Annals of Mathematical Studies, No. 28, Princeton University Press, Princeton, NJ.

Shapley, L. S., 1962. 'Values of Games with Infinitely Many Players', in Maschler, M., ed., *Recent Advances in Game Theory*, Princeton University Press, Princeton, NJ.

Shapley, L. S., 1967. 'On Balanced Sets and Cores', *Naval Research Logistics Quarterly* 14: 453–60.

Shavell, S., 1979. 'Risk Sharing and Incentives in the Principal and Agent Relationship', *Bell Journal of Economics* 10: 55–73.

Shubik, M., 1959. 'Edgeworth Market Games', in Tucker A. W. and Luce, D. R., eds, *Contributions to the Theory of Games*, Vol. IV, Annals of Mathematical Studies, No. 40, Princeton University Press, Princeton, NJ.

Shubik, M., 1971. 'The Dollar Auction Game: A Paradox in Noncooperative Behavior and Escalation', *Journal of Conflict Resolution* 15: 109–11.

Shubik, M., 1982. *Game Theory in the Social Sciences*, MIT Press, Cambridge, MA.

Shy, O., 1995. *Industrial Organization. Theory and Applications*, MIT Press, Cambridge, MA.

Shy, O., 2001. *The Economics of Network Industries*, Cambridge University Press, Cambridge.

Siegfried, T., 2006. *A Beautiful Math. John Nash, Game Theory, and the Modern Quest for a Code of Nature*, Joseph Henry Press, Washington, DC.

Skaperdas, S. and Syropoulos, C., 2001. 'Guns, Butter, and Openness: On the Relationship Between Security and Trade', *American Economic Review* 91: 353–7.

Slutsky, S. M., 1975. 'Abstentions and Majority Equilibrium', *Journal of Economic Theory* 1: 292–304.

Smith, J., 1998. *The Cold War, 1945–1991*, Blackwell, Oxford.

Snidal, D., 1985. 'Coordination versus Prisoner's Dilemma: Implications for International Cooperation and Regimes', *American Political Science Review* 79: 923–42.

Solow, R., 1956. 'A Contribution to the Theory of Economic Growth', *Quarterly Journal of Economics* 70: 65–94.

Spence, A. M., 1973. 'Job Market Signalling', *Quarterly Journal of Economics* 87: 355–74.

Stern, N., 2007. *The Economics of Climate Change: The Stern Review*, Cambridge University Press, Cambridge.

Stern, N., 2009. *A Blueprint for a Safer Planet. How to Manage Climate Change and Create a New Era of Progress and Prosperity*, Random House, New York.

Stiglitz, J., 1985. 'Information and Economic Analysis: A Perspective', *Economic Journal* 95: 21–41.

Stiglitz, J. and Weiss, A., 1981. 'Credit Rationing in Markets with Imperfect Information', *American Economic Review* 71: 393–410.

Swan, T. W., 1956. 'Economic Growth and Capital Accumulation', *Economic Record* 32: 334–61.

Sylos-Labini, P., 1962. *Oligopoly and Technical Progress*, Harvard University Press, Cambridge, MA.

Tannenwald, N., 1999. 'The Nuclear Taboo: The United States and the Normative Basis of Nuclear Non-Use', *International Organization* 53: 433–68.

Tirole, J., 1988. *The Theory of Industrial Organization*, MIT Press, Cambridge, MA.

Tisdell, C., 2009. *Resource and Environmental Economics. Modern Issues and Applications*, World Scientific, New York.

van Damme, E., 1989. 'Stable Equilibria and Forward Induction', *Journal of Economic Theory* 48: 476–96.

van Damme, E., 1991. *Stability and Perfection of Nash Equilibria*, Springer, Heidelberg.

van Damme, E. and Weibull, J., 1995. 'Equilibrium in Strategic Interaction: The Contributions of John C. Harsanyi, John F. Nash and Reinhard Selten', *Scandinavian Journal of Economics* 97: 15–40.

Varian, H., 1992. *Microeconomic Analysis*, 3rd edn, Norton, New York.

Vasquez, J. A. and Elman, C., eds, 2000. *Realism and the Balancing of Power*, MIT Press, Cambridge, MA.

Vickrey, W., 1962. 'Auctions and Bidding Games', in *Recent Advances in Game Theory*, pp. 15–27, Princeton University Conference Series, Princeton University Press, Princeton, NJ.

von Neumann, J., 1928. 'Zur Theorie der Gesellschaftsspiele', *Mathematische Annalen* 100: 295–320.

von Neumann, J., 1932. *Mathematische Grundlagen der Quantenmechanik*, Springer, Berlin. [See also: *Mathematical Foundations of Quantum Mechanics*, Princeton University Press, Princeton, NJ, 1955.]

von Neumann, J. and Morgenstern, O., 1944. *Theory of Games and Economic Behavior*, Princeton University Press, Princeton, NJ.

Wagner, R. H., 1991. 'Nuclear Deterrence, Counterforce Strategies, and the Incentive to Strike First', *American Political Science Review* 85: 727–49.

Wagner, R. H., 2004. 'Bargaining, War, and Alliances', *Conflict Management and Peace Science* 21: 215–31.

Walt, S. M., 1987. *The Origins of Alliances*, Cornell University Press, Ithaca, NY.

Waltz, K., 1979. *Theory of International Politics*, Addison Wesley, Reading, MA.

Weber, S., 1997. 'Entry Deterrence in Electoral Spatial Competition', *Social Choice and Welfare* 15: 31–56.

Weintraub, R., ed., 1992. 'Toward a History of Game Theory', *History of Political Economy*, 24(Supplement).

Williamson, O. E., 1975. *Markets and Hierarchies. Analysis and Antitrust Implications*, Macmillan, New York.

Wittman, D., 1977. 'Candidates with Policy Preferences: A Dynamic Model', *Journal of Economic Theory* 14: 180–9.

Wittman, D., 1983. 'Candidate Motivation: A Synthesis of Alternative Theories', *American Political Science Review* 77: 142–57.

Wittman, D., 1990. 'Spatial Strategies When Candidates Have Policy Preferences', in Enelow, J. M. and Hinich, M. J., eds, *Advances in the Spatial Theory of Voting*, Cambridge University Press, Cambridge.

Yang, C.-L., Yue, C.-S. J. and Yu, I.-T., 2007. 'The Rise of Cooperation in Correlated Matching Prisoners Dilemma: An Experiment', *Experimental Economics* 10: 3–20.

Yarbrough, B. V. and Yarbrough, R. M., 1986. 'Reciprocity, Bilateralism, and Economic 'Hostages': Self-Enforcing Agreements in International Trade', *International Studies Quarterly* 30: 7–21.

Yarbrough, B. V. and Yarbrough, R. M., 1987. 'Cooperation in the Liberalization of International Trade: After Hegemony, What?', *International Organization* 41: 1–26.

Zermelo, E., 1913. 'Über eine Anwendung der Mengenlehre auf die Theorie des Schachspiels', *Proceedings of the Fifth Congress of Mathematicians*, Vol. 2, pp. 501–4.

Index

adverse selection, 130, 135, 142
advertising, xiii, 15, 47, 49, 50, 56, 57, 85, 88,
    93, 135, 142
agency model, 130–135
antitrust, 53, 73, 76, 79
arms race, xi, 7, 12, 101, 102, 118, 146
Arrow's impossibility theorem, 86, 87
auction, 50–52, 58, 97, 101, 102
axiomatic approach, 87, 150–152, 156

backward induction, 2, 4, 32–34, 36, 37, 42, 72,
    73, 95, 108, 117, 127, 129, 145
    and forward induction, 144
bargaining, 4, 5, 149–154
battle of the sexes game, 29
Bertrand, 48
    equilibrium, 48, 49, 77, 79, 153
    game, 48, 50, 55, 74, 75, 77, 78
    model, 167
    paradox, 49, 74, 75, 79

centipede game, 37
chain store paradox, 82, 83
characteristic function, 155, 156
chicken game, 29, 30
coalition, 154–156, 158, 160–162
    grand, 155, 156, 158
    proper, 155, 156, 162, 163
Cold War, xi–xiii, 6, 8, 9, 12, 87, 97, 102, 103,
    107, 108, 119
collusion, 69, 73, 75–80, 82, 152
Condorcet paradox, xiii, 86
cooperation, 149, 150, 154, 156
    in experiments, 82
    in international relations, 113
coordination game, xiii, 30, 32, 39, 41, 60, 109,
    110, 127, 145, 146
core, 5, 154, 156–158, 160
Cournot, 3
    game, 48, 50, 74, 75

defensive weapons, 106, 108, 109
disagreement point, 150, 151, 153, 154

dominant strategy, xiii, 4, 26
    equilibrium, 26, 27
    strictly, 26
    weakly, 26
Downs model, 88, 90
    with platform stickiness, 91

eBay, 50
egalitarian solution, 154
elections, xiii, 85, 86, 89, 91, 93–95
entry and entry barriers, xiii, 56, 58, 59, 82–84
    with incomplete information, 137–141
environmental economics, xiii, 7, 59, 66–68, 162
experimental approach, 7, 37, 38, 41, 44, 82
extensive form game, xii, 4, 16–19

fiscal policy, 63, 66
focal point equilibrium, 26, 31, 32
folk theorem, xiii, 4, 5, 73, 75, 76, 78, 80
    perfect, 77, 80, 115, 124, 126, 143
forward induction, 129, 144–147
    and backward induction, 145
    and iterated dominance, 146
free trade, 113–116
free-riding, 4, 28, 114, 162, 163

hegemony, xiii, 12, 14, 113, 119, 120
hidden action, 130–132
hidden information, 130–132
Hotelling model, 54, 56, 57, 90

imputation, 155, 156
incentive compatibility constraint, 133
information, xiii, 2, 15, 21, 25, 27, 29, 37
    asymmetric, 15, 129, 130, 135, 142
    complete, 4, 15, 29
    imperfect, 15, 16, 29
    incomplete, 4, 15, 129, 136
    perfect, 4, 15, 16, 33, 37
    symmetric, 15

innovation, 47, 49, 50, 56, 57
  and entry, 58
  and market power, 50, 53
  process, 50
  product, 50
iterated deletion, 29, 31

Kuhn tree, 4, 32, 95, 101, 104, 138

macroeconomic policy coordination, 66
MAD, 103
matching pennies game, 39–41
median voter, 89–93
  theorem, 90
minimax equilibrium, 2, 3, 22
mixed strategy, 14, 40
monetary policy, 63, 66
moral hazard, 130–132, 142
multiplicity of equilibria, 26, 30, 32

Nash bargaining solution, xiii, 5, 150, 151, 153
Nash equilibrium, xiii, 3, 4, 24, 25
  in mixed strategies, 25, 39–41, 100, 101, 105, 110
  refinements of, 26
Nash product, 151
Nash programme, 4, 150
network externality, 47, 59–62
new trade theory, 114
node, 16, 17
normal form game, 16

one-shot game, 15

Pareto efficiency, 4, 23, 28, 30–32, 36, 39, 54, 63, 66, 68–70, 72, 73, 75, 76, 82, 92, 106, 109, 114, 115, 117–119, 123, 124, 127, 145, 149, 151, 152, 154, 156, 159
participation constraint, 133
payoff, 14
perfect Bayesian equilibrium, 4, 136, 139
pooling equilibrium, 137, 141, 144, 148
prisoners' dilemma, xiii, 4, 7, 27, 28, 36, 57, 64, 66, 68, 69, 72, 73, 75–78, 80–82, 92, 94, 102, 109, 114–116, 118, 119, 123–125
product differentiation, 47, 54–56, 153
product quality, 52, 60, 61, 130, 135, 142, 143
protectionism, 113–115

public goods, 63, 64, 120
punishment, 4, 77–80, 115

quality premium, 142
quantum physics, 43

RAND Corporation, xii, 3, 6–8, 103
repeated games, xiii, 4, 15, 41, 69, 70, 72, 73, 75–80, 82, 83, 115, 116, 124, 126, 136, 139, 140, 142
reputation, 129, 136, 141–144
risk dominance, 26, 39, 110
  in mixed strategies, 41

Schrödinger's paradox, 43
separating equilibrium, 136, 137, 140, 141, 144, 148
Shapley value, 156, 157, 159, 160
signalling, 142, 144–146
singleton, 16, 17, 19
stag hunt game, 30
standardization, 47, 59–61, 63
Strategic Defense Initiative, 108
strategic form game, xii, 16
strictly competitive games, 14, 22, 99
subgame, 17
subgame perfect equilibrium, xiii, 4, 17, 26, 32–34, 36–38
supergame, 4, 15, 70–73, 76, 77, 80, 81, 83, 84, 115, 124, 126, 142, 143

time discounting, 7, 70, 71, 77, 80
tit-for-tat strategies, 77, 80, 81
tragedy of commons, 66–68
trigger strategies, 77, 115, 124

ultimatum game, 37, 38
utility, 2, 60, 61, 64, 88, 91, 95, 114, 120, 121, 131, 133, 143, 150, 151, 155, 161
  expected, 44, 133
  reservation, 131, 133
  transferable, 153, 155

variable-sum games, 14, 22, 23, 27

Zermelo's theorem, 2
zero-sum games, 14, 22

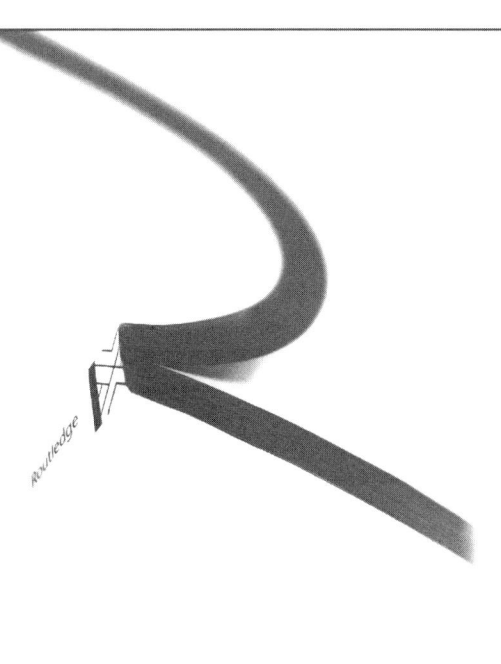